Analecta Gregoriana 315

GENNARO AULETTA

INTEGRATED COGNITIVE STRATEGIES IN A CHANGING WORLD

in collaboration with I. Colagè, P. D'Ambrosio, L. Torcal

GREGORIAN & BIBLICAL PRESS

We wish to warmly thank for help and suggestions Mgr. Armand Puig, Dr. Ignacio Silva, Prof. João J. Vila-Cha, SJ, Prof. Sergio P. Bonanni, Fr. Hans Zollner, SJ, and Fr. Nicola Gobbi, SJ. We also thank the student C. Andrews for linguistic revision.

Progetto grafico di copertina: Serena Aureli

Impaginazione: Lisanti Srl - Roma

ISBN 978-88-7839-**207**-6

Perhaps someone will say that we have not understood Aristotle and that on this account we have not agreed with what he said or that we contradict him on points of truth in some matter. To him we say that whoever believes that Aristotle was a god ought also to believe that he never erred. However, if one believes him to be a man, than without doubt he could err just as we can do.

Albert the Great, Liber Phys., VIII, tr. 1, c. 14.

PREFACE
A Brief History of the Book

Collaborating to the preparation of this book has been a crucial intellectual experience that decisively contributed to our formation. It all began in the second week of July 2009, when we were attending the 3rd edition of the Summer School on Evolution held, as usual since 2007, at the Monastery of Poblet (Catalonia) under the guidance of Professor Gennaro Auletta. He had scheduled a private meeting with the three of us after the second morning session of the School. We just did not know what to expect from that, and it had been amazing to realize that he wanted us to join him for a paper addressed at sketching the main guidelines of a renewed philosophy of nature consistent with modern scientific knowledge. Indeed, we were well aware of Gennaro's great interest in the field not only for his professional experience but also because he used to frame the work of his group along pathways leading to the critical assessment of primary philosophical issues concerning the natural world. However, the idea of trying to write down a paper with him on the foundations of a renewed philosophy of nature struck us a little at first.

Gennaro introduced us to his main ideas about the matter and in particular to three general philosophical principles that, he argued, might have been plausibly acknowledged from what science, in different areas of investigation, show us about the very constitution of the natural world. Moreover, he hinted at the possibility of a theological interpretation of those principles according to the Thomistic perspective, thus envisaging an actual contribution to the philosophically-mediated dialogue between natural science and theology of Creation. A quite long and lively discussion followed, each of us taking so many notes as possible. At the end of that first brainstorming, we came to understand the project a little better and enthusiastically endorsed it. Some days later, Gennaro started to write down the text, while the three of us put together all our notes so to help in outlining some particular issues. After some common working sessions, a first draft of more than 30 pages was produced in October, to which several sections and subsections were added during the following months. By the end of 2009, we were working on a draft of about

100 pages. Already after this first period of work, the feasibility of the project and its promise became clearer to our eyes and the contours of the original view, presented to us during the Summer School at Poblet, quite delineated. The more we went on in elaborating the mentioned insights into the three principles, the more their relevance in grasping the fundamental lessons coming from several scientific disciplines appeared in a proper light. Moreover, the perspective of a possible theological interpretation of the principles took form cogently.

If the three philosophical principles may still represent the conceptual core of the present volume, those early drafts of the manuscript have been so drastically changed that one can barely find a trace. Even the original idea of writing a paper was discarded as Gennaro more and more headed into the direction of a programmatic essay delineating, on the one hand, the main tenets of our working team and, on the other, the ways to conceive the strategies to favour a cognitive enterprise encompassing science, philosophy and theology as well. In such a fascinating wider perspective, the result obtained at the end of 2009 merely appeared as a starting point requiring further study and research: the parts dedicated to epistemology (present Chapter 2) and to philosophy of nature (present Chapter 3) had to be deepened and enlarged, while the theological part (present Chapter 4) had to take into account the works of some of the main contemporary theologians who dealt with issues concerning the scientific knowledge. In brief, it was mainly thanks to Gennaro's unceasing efforts that the manuscript has been progressively enriched up to the 200 pages draft produced by the end of 2010, also including a brand new part focussed on anthropology (present Chapter 5). In such a process, we mainly tried to follow his intellectual advancements suggesting, as far as we could, possible ways to further exemplify the appropriateness of the three principles or to formulate some parts of the argumentation in convenient terms. This gave raise to animated debates among the four of us leading to rethink previous views, rectify formulations and restlessly check the correctness of the proposed conclusions in terms of both current scientific findings and philosophical rigour. Further updating, reformulations and bibliographical references were inserted in the revised version finally submitted to the publisher in June 2011.

Working on this project, the three of us had the opportunity to partake to a research engagement going far beyond the specialized area of interest within which doctoral studies are usually confined. We have been stimulated and challenged by Gennaro's outstanding capability of posing radical questions, of going at the heart of significant problems, and of critically evaluating possible solutions at the same time figuring out possible connections between apparently unrelated issues. The limited contribution that we were able to provide is essentially related to the discussions we had during the team working sessions, which brought about developments of original ideas, that is, of the actual backbone of the volume. Besides, we worked hard in surveying papers and books helpful for the argumentations to be developed as well as in the drafting, with the aim of offering an open hand to the interested reader.

As Gennaro nicely said to us during one of the last working sessions, he viewed the process of writing this book as a prolonged high-quality research-seminar in which the "foundations" of our team were further refined and expressed in a document that might be under many respects considered as a manifest of our long-term research project. We could not agree more on that, and we feel privileged to have taken part to such an enterprise.

To the reader, we would like to synthesize a general lesson that we have drawn from this experience. Philosophy should listen, especially today, to what science has to say about the natural world. But it should not listen passively; rather, it should re-discover its original attitude, which is at the same time critical and future-oriented. On the one hand, philosophy is called to carefully assess the main intellectual trends of the present times also in the light of past traditions. In doing this, it may also significantly contribute in defining the strategic lines along which the human cognitive enterprise is to be fostered. Past philosophers, from Plato and Aristotle to the Moderns and beyond, always have fundamental lessons to convey, which, however, need to be put in dialogue with the contemporary cultural environment (in its technological, scientific, anthropological and even theological facets), thus becoming actualized and able to favour new research strategies and intellectual contexts for future generations. This seems to be one of the most pressing priorities for the present-day philosopher as he/she faces the deep changes affecting the cultural world of our times.

In a way, we have been the first "readers" of this book. We took from it much more than what we were able to give. Thus, we would like to express all our sincere gratitude to Gennaro for having involved us in a project that turned out to be extremely relevant for our intellectual growth. Indeed, it has been not only a matter of dealing with so many different topics but also, and above all, of learning how *to do* philosophy in a professional way.

Rome, September the 12th, 2011
Ivan Colagè, Paolo D'Ambrosio and Lluc Torcal

TABLE OF CONTENTS

1

INTRODUCTION

1.1 Theories, Approaches, and Traditions

There are many ways to understand the relations among science, theology and philosophy. Apart from negative standpoints (some of which will be treated in the following), an important approach to this matter is the traditional attempt at highlighting the compatibility or at least the non-contradiction between the Holy Scriptures and science. The fathers of modern science had already felt such urgency. We may recall here the work of Galilei as one of the most representative[1]. Recently, remarkable attempts have been made at showing that there is no contradiction between the Scriptures and theory of evolution[2]. Another interesting way to understand the relations between science, philosophy and theology is represented by those who deal with this problem within the framework of fundamental theology with the aim of updating the latter field[3]. Finally, there is a growing interest by many scientists for traditional philosophical and theological issues as testified today by many books of famous scientists dealing with ultimate questions;[4] a great variety of positions are displayed, some favourable to a religious approach, others less so.

In this book we shall be more interested in dealing with the empirical results of the natural sciences as well as with their general solutions and theoretical constructions, and in relating them to specific fields of both philosophy and theology, namely, philosophy and theology *of nature*. The rapid development of science

---

[1] Galilei 1615. There is however a long tradition behind this and we can take Averroes' work as one of the first attempts at reconciling faith with reason (this philosopher is often charged to be a theoretician of double truth, which is not true: he only warned people to be careful in communicating deeper intrepretations of religious matter to uninstructed people since this could lead to several misunderstandings: see Averroes *DD*, p. 64).

[2] Alexander 2008.

[3] Tanzella-Nitti 1994; 2002; 2008; 2009.

[4] As also pointed out by Benedict XVI (2011).

nowadays and the ongoing paradigmatic changes (particularly involving biology and neuroscience), make this attempt a very urgent one but also much more feasible than in the past, as we shall demonstrate. On the other hand, we are interested in understanding what could be the contribution of philosophy and theology to a very dynamic stage of scientific research.

We are deeply convinced that the natural sciences may play today a similar role for Christian theology to the one that Platonism offered to Patristic theology and Aristotelism to Medieval theology[5]. Therefore, the present work should be understood as an attempt at contributing to a renewal of philosophy and theology according to the methodological guidelines of Patristic and Scholastic approaches[6]. On the other hand, in the past, philosophy and theology have been very helpful in fostering scientific enterprises. Summing up, our investigation enters into the fields of science, philosophy and theology in order to promote a dynamic interchange, possibly providing new insights and cognitive strategies. As we shall argue in the following, the interactions between science, philosophy and theology are not only pragmatically helpful, but moreover are actually rooted in a common general cognitive framework. Indeed, each of these fields, although applying different methodologies and existing in full autonomy, pursues a search for truth, and knowledge in each of them ultimately means rationally justified belief.

Nevertheless, it is also very important to understand what the specificities of these three different fields are from the start. The dynamic result of scientific activity is represented by *theories*. Thus, the scientific enterprise is characterized by a spirit of innovation which often leads to a certain uneasiness towards any established point of view and therefore towards cultural traditions as well, which consist in consolidated and integrated perspectives. For this reason, science cannot constitute a culture by itself[7], which demands the ability to integrate different dimensions of human life on the basis of a social consensus. Theology, on the contrary, is tightly connected with historical revelation and religious *traditions*, which are core elements of cultures, and therefore, its cognitive import is also rooted in traditions[8]. As we shall see below, this does not imply that theology is

---

[5] McGrath 2001-2003, v.1, p. 7.

[6] John Paul II 1988.

[7] Dyson 2008.

[8] Unfortunately, many scientists consider religious traditions (even mythological ones) as kinds of pre-scientific theories of the world (Hawking-Mlodinow 2010, pp. 15, 87, 123, 149), which is not quite accurate. We shall come back to this in the following.

devoid of rational content, rather it should be understood as a rational purification of traditions. Indeed, it is only religion that is necessarily connected with specific customs, while theology, although still depending on particular religious traditions, also has a universal aspiration. This is why Christianity has, since its beginnings, emphasized its truth against the pagan customs that have a mythical basis[9]. Philosophy, on the other hand, is not expressed through particular theories (like science), neither is it articulated in historical traditions in the same sense as theology, even if historical schools of thought can be acknowledged. Philosophy deals rather with some general characters of our world and of knowledge about this world as well as with the implications and consequences for humanity, and, in this sense, the different solutions that have been expressed throughout its history (like Platonism, Aristotelism, Idealism) could be called *approaches*. The term "approach" means that philosophy is not so much articulated in terms of positive proposals (like scientific theories are) but rather consists of a *critical* activity and in its ability to formulate general research strategies that somehow influence or determine the general framework in which scientific research is developed.

1.2 Two strategies of research

Given these considerations, it is obvious that theology cannot have a *direct* relation with science, and therefore cannot incorporate any scientific theory in its framework. This even sometimes causes troubles and difficulties when people try to translate theological language into scientific language or vice-versa. We think that a reciprocally profitable interchange between theology and science requires more complex relations, what also involves the critical contribution of philosophical approaches. To better understand this point, let us consider some relevant historical examples of the way in which theology and philosophy may be crucial for science in the light of a new concept, namely that of *research strategy*. Kuhn formulated the idea of a research paradigm[10] meaning the general framework in which some scientific researches are developed, like Ptolemaic astronomy, Newtonian dynamics, corpuscular optics, and so on. Here, the term *research strategy* is more general, referring to the set of methodologies and orientations that determine the research in *several* fields over a very long time. There have been essentially two research strategies along the history of the Western world. The first was expressed in the Ancient

---

[9] Pope Benedict XVI, www.zenit.org/article-9138?|=italian.
[10] Kuhn 1962

world, especially through Plato's philosophy of forms[11]. The second has been expressed in modern times. The former strategy gave rise to relevant results such as the concept of *logos* and a rational language as well as to a science that was essentially formal and descriptive. We recall the major scientific fields that came out of this approach: Euclidean geometry, basic arithmetics, Ptolemaic astronomy, and Archimedes' lever statics and hydrostatics[12]. It is true that Aristotle tried to give another direction to philosophy introducing a new sensibility for the natural world (even if his contribution to logic was also very relevant). In particular, he tried to consider the dynamics of natural processes and introduced the idea that causal explanations were relevant to knowledge; moreover he developed a philosophical examination of the concept of causality. However, the empirical data at that time were largely insufficient to support this point of view that remained essentially unripe and could not be translated into a concrete scientific research program. Now, it is difficult to understand the philosophical approach of Plato and his research strategy without a previous purification of the cultural (traditional) environment in which Greek philosophers lived and thought. This was provided by a number of poets and thinkers who founded a new interpretation of religion based on a genealogical mythology of Olympus' divinities. This was especially due to the work of Hesiod[13]. These gods were still part of a mythological understanding of reality, but expressed the characters of heavenly things, like beauty and self-satisfaction, without which it is difficult to think of a world of pure forms.

With regards to the second qualified research strategy, applied in the modern era, A. Koyré[14] rightly stressed its Platonic-mathematical character. Indeed, mathematical developments like the differential calculus were crucial. However, this only represents a part of the problem, the other is the search for mechanical (and therefore also causal) explanations. This is rather connected with the Aristotelian approach, as acknowledged by R. Hall[15], despite the fact that modern science essentially rejected Aristotelism due to the already mentioned unripeness of his nat-

---

[11] Lindberg 1992

[12] The astonishment towards the axiomatic and formal character of Archimedes' science shown in Hawking-Mlodinow 2010, p. 20, portrays a deep misunderstanding about the character of ancient science.

[13] Clagett 1955, p. 22; Salucci 2011, § 1.5.

[14] Koyré 1966.

[15] Hall 1954, p. 188.

ural philosophy. Certainly, also the book *Whatever Arsistotle has said is false,* written in an elegant Latin by the French humanist Petrus Ramus contributed to discredit Aristotle's philosophy, although it showed little understanding of logic and epistemology. The synthesis of Platonism and Aristotelism was provided in the Middle Ages (relying on the immense intellectual work carried out in the Islamic countries), especially through the work of St Thomas, who, on the one hand, accepted Platonism (especially St Augustine's reinterpretation of the world of Platonic forms as God's mind) and, on the other, with his distinction between the Primary Cause and secondary causes, prepared the path for naturalistic explanations based on causality (and therefore no longer being purely descriptive).[16] If there is something in common to all philosophers (like Hobbes, Locke, Galilei, and many others) who were grounding the new natural science of the Modern Ages, it is the search for mechanisms and therefore the refusal of essences as the explanation for natural phenomena[17]. Often, this is taken as a standpoint against Middle-Age theology. It is, however, a deep misunderstanding since the distinction quoted above between the Primary cause and secondary causes is precisely the background of a scientific research strategy that is founded on naturalism. According to this point of view, we may possibly arrive at an understanding of certain essences through empirical research, but *not the other way around.* In other words, we cannot predetermine our experience on the basis of a priori knowledge. To do so would run against St Thomas' other foundational distinction between what is known in itself (*per se*) and what is known to us (*quoad nos*)[18]. If we had the cognitive power to grasp ideas *per se*, we should proceed from the "exposition of God, as he is in his eternal being before the creation of nature and of a finite spirit", as Hegel boldly writes in the Introduction to his *Science of Logic*[19]. In other words, we should be able to expose the "self-movement of the absolute Idea [...] as the original Word"[20],

---

[16] Duhem 1913-1959. It is worth mentioning that William of Ockham's nominalism was characterized by the rejection of causal explanations (a standpoint also exemplified by Hume's criticism of causality). This proposal was not generally accepted and this is also true for some of the concrete solutions proposed by Ockham himself, although he raised interesting ideas opening new research paths (see Pedersen 2007, pp. 200-201).

[17] Galilei 1612. Locke 1689, Book II, Ch. 23; Book IV, Ch. 6, §§4-6 and Ch. 12 §§9-12. See Cassirer 1910.

[18] Aquinas *S. Th.*, I, q. 2, a. 1. See also Aristotle, *An. Post.,* 71b34-72a5.

[19] Hegel 1833, v. I, p. 31.

[20] Hegel 1833, v. II, p. 485.

which is, according to Hegel, the science of the divine concept[21]. It is also worth mentioning that some of the conclusions of the modern empirical research were already worked out during the Middle Ages and often with an overt criticism to Aristotle's physics. We recall here the studies started by John Philoponus and developed by John Buridan, Robert Grosseteste, Nicole Oresme and many others[22]. Here there is another important commonality between scholasticism and science in the Modern Ages: the spirit of a minute, systematic, extensive, strict research, promoted in a collective style, which was so well understood by Peirce[23]. Therefore, this Middle-Age theological and philosophical synthesis may be seen as preparing the path to the Modern-Age formulation of the new mechanist research strategy, even if, ironically, modern science has subsequently not only forgotten its theological-philosophical background but also vigorously separated itself from philosophy when it seemed to be in a state of dissonance with its strategies (as it was the case for German Idealism or the so called Dilthey's historicism).[24] It is quite possible that this historical fracture is explainable, at least in part, by the new theological understanding brought by the Reformation: most of the leaders of the reformed Churches stressed that scholasticism abused human reason by minimizing the weakness of the latter, determined by original sin[25]. For this reason, they showed mostly a sceptical attitude towards natural sciences, although with the side effect that many Modern-Age scientists considered themselves to be innovators relative to the Middle Ages. Most of the reformed theologians made a partial exception to their sceptiscism only for the understanding of mathematical truths. This could explain in part the Platonic style, characterizing the early stages of modern science in particular, although causal explanations became eventually more and more important as science continued further and further. Another important consideration is that changes occurring in scientific strategies could not completely wash out methodologies and results of the previous ones. In any case, such an important tradition, with all its contradictions and shifts in understand-

---

[21] Hegel 1833, v. II, p. 505.

[22] See Pedersen 2007, pp. 119-120 and 172-177.

[23] Peirce CP, 1.32-33. For this reason we agree with Harrison (2007, p. 249) that "to think small" is not a prerogative of modern scientists alone.

[24] Among the great modern minds only Leibniz tried to combine modern science with Aristotelian tradition (and this quite unsuccessfully, for reasons that in part will be explained below).

[25] Harrison 2007, pp. 54-61 and 89-107.

ing, fully justifies the role of Christian theology when in the past the relations with science and philosophy were of crucial relevance.

Summing up, we have shown that, for the two fundamental research strategies of Western science, both philosophy and theology have played a major role (although for the ancient world it was rather a sort of philosophical theology). We believe this is true not only for the past but also for the future, and we would even go as far as saying that a research strategy is established only when there is a major integration between scientific, philosophical and theological perspectives. Is not our post-modern era already facing such a situation?

1.3 The Current Change

There are reasons to believe that a new scientific strategy is emerging[26]. Since the modern mechanist research strategy was born and consolidated in physics (the queen of all sciences), it is natural to expect that the first problems occur in this field. Indeed, this is exactly what happened. We recall here the following developments:

1. Thermodynamics. This discipline showed for the first time that engines are not self-sustaining: they do not work eternally (they undergo disruption) and could not have emerged spontaneously from a mechanical world.
2. Classical electromagnetism, developed from a set of ideas originally due to Leibniz and Boscovich[27]. Einstein[28] thought that the concept of field introduced by this theory was not fully compatible with the mechanical action of forces. Classical mechanics tried to encompass these developments by introducing the fiction of a fixed field-producing charge and a moving one, and so treating the electromagnetic field in a way that was compatible with mechanics (here one particle acts on the other). This, however, runs into certain difficulties[29] and could not provide a general model of fields.
3. Special and General Relativity. These two theories can be understood as a generalization of the concept of field.
4. Quantum mechanics. We shall show throughout this book what its novelty is in relation to classical mechanics. We shall focus on two concepts, namely random events and correlations.

---

[26] Auletta 2010.
[27] Leibniz 1686a. Boscovich 1758.
[28] Einstein MW, pp. 160-161.
[29] Margenau 1950, pp. 36-39.

5. Theory of complex systems. This theory will also be discussed in the following. The reasons for its incompatibility with classical mechanics are similar to those of quantum mechanics (amplification of random fluctuations and long-ranging correlations).

As a matter of fact, despite the huge (and even very general) relevance of these results, they have shown to be insufficient at changing the framework of mechanist science to date, which continues to thrive, having given rise to fields like molecular biology and nanoscience. We also mention that classical mechanics has even been able to absorb elements of the theory of complex systems (for instance with the Kolmogorov-Arnold-Moser theorem[30]).

To understand this incredible powerfulness of classical mechanics, let us consider the following point. Also Newton's gravitational theory was felt as being incompatible with a classical-mechanical framework, because it was understood as an action-at-a-distance. This was also Newton's personal opinion[31] but especially that of the fathers of the new classical-mechanical synthesis at the beginning of the 19th century, especially D'Alembert[32]. Nevertheless, the latter theory was able later on to encompass the law of gravitation in its building as a special case of motion[33] (see Appendix). These examples are impressive and can give the reader an idea of what we mean by the notion of scientific *research strategy*.

So, the aforementioned five results should rather be understood as the preliminary steps that shall enable the formulation of a new synthesis rather than as a true conceptual reversal. This is likely to come in the future, and will bring with it the ability to find new typologies of explanation. However, we expect that such a change cannot come from physics alone (although anticipated in such a field) but requires totally different kinds of problems like those arising in disciplines like biology and neuroscience. If we are right, such a historical change will necessarily involve philosophy and theology. It is only philosophy that can provide for a correct and general formulation of a new research strategy (after that of antiquity and that of the Modern Ages). We shall propose three heuristic principles that can help this conceptual effort in establishing a very general framework and a new integration of humanism and naturalism based on the critical interpolation of

---

[30] Arnold 1978, pp. 404-415.
[31] Newton 1687. See also Koyré 1957.
[32] Rossi 1973-1974. See also Leibniz 1689a.
[33] Landau–Liftshifts 1976, Sec 10. The first idea in this sense is again due to Boscovich.

three fundamental philosophical approaches: Platonism, Aristotelism, and Idealism. It is worth recalling the first attempt made in this direction by Maréchal, who, still essentially relying on Kant's philosophy, tried to articulate Idealism with the Thomistic approach that already integrated Platonism and Aristotelism[34]. Such a synthesis, however, requires a proper cultural humus. Here, theology, as a purification of the cultural background on which philosophy and science work, can be very helpful. It is impossible to foresee the form that this background will take, even if we shall present some possible ideas in the following. Our Western countries sometimes appear to be culturally exhausted in the sense that there seems to be a fundamental inability to bring about innovative ideas. When science forgets its philosophical roots and segregates itself from theological insights, it appears to be useful for applied research but much less apt at producing new ideas and fertile solutions[35]. Also, philosophy without a connection with science and theology seems to be trapped within an empty relativism as it is often the case in continental Europe. It is possible that this new cultural humus will emerge in certain developing countries and not in Western societies. In any case, this problem should not leave Christian theologians, philosophers, and scientists indifferent. Summarizing, the next scientific turn is still ahead but the process has already begun in its initial steps; thus, it would be of maximal interest for these subjects to facilitate this process because it will represent a major accomplishment in the cognitive journey of humankind.

1.4 First steps

The present book tries to develop an integrated view of knowledge[36] in a dynamic sense. We shall consider successes, attempts, failures, and problems according to past experience: we shall introduce many examples drawn from the history of science, constituting a sort of empirical background of our investigation. However, as mentioned, we shall also consider present stimuli coming from scientific research and finally try to trace possible perspectives, especially sketching what the general character of a new research strategy could be.

---

[34] Maréchal 1926.

[35] Einstein said that whenever the religious confidence "in the rational nature of reality" is absent, "science degenerates into uninspired empiricism" (quoted in Jammer 1999, p. 120).

[36] *Fides et Ratio*, n. 34. See Tanzella-Nitti 2009, Ch 8, for a discussion of some problems.

In spite of the abstractness of the subject of the present book, we think that the questions we are dealing with are very concrete and have a pragmatic import as they are connected with empirical developments in science as well as with some needs arising from current philosophical and theological perspectives.

In other words, we are trying to frame the envisaged cognitive enterprise in the context of a very important transformation occurring from the *inside* of scientific inquiry. We are fully aware that this process is currently in progress so that, at the moment, it is not possible to fully understand what exact configuration the natural sciences will eventually assume. Moreover, we are also aware that an integrated cognitive enterprise of the kind we are wishing requires an ensemble of competencies that cannot be mastered by a single individual, and neither, perhaps, could be found in a single institution. Furthermore, the required joint contribution of different perspectives is very difficult to achieve. In any case, the present contribution should be understood as a step into this direction whose purpose is to contribute to the formulation of a new philosophy of nature that captures the interesting and significant changes of our scientific image of natural reality that are arising right now.

2

AN EPISTEMOLOGICAL ASSESSMENT

When dealing with the issue of understanding our universe, *natural science,* in both its empirical and theoretical dimensions, is the basic level of knowledge: in our opinion, neither philosophy of nature nor a theological reflection on the natural world could be currently developed without a pertinent reference to scientific contents and procedures[1]. In the present chapter, we shall focus on some relevant epistemological and methodological aspects of the three fields here involved. This will bring us to solve, and sometimes to dissolve, some problems affecting the interaction among them, as well as to enucleate the *real* problems to be faced in achieving a truly integrated cognitive enterprise.

2.1 Sciences of Nature

2.1.1 *Scientific Theories*

Broadly speaking, the first aim of science is to make natural phenomena *predictable* and *controllable.* Successful interactions with nature and predictions about it are possible thanks to *regularities*: any true advancement in science basically requires finding out some new (and maybe unexpected) regularities[2], which are usually stated in *laws.* However, laws, if isolated and unrelated, are of little help when dealing with the complex reality that surrounds us. They must be harmonized and articulated in a conceptual framework commonly meant as a scientific *theory.* Therefore, theories cannot and should not be reduced to mere aggregates of laws (or to collections of facts); they are creatively elaborated to *explain* regularities[3], i.e., (i) to single out the experimental or natural-spontaneous contexts in which these regularities happen, (ii) to provide elements from which they may be derived or justified, and (iii) to enquire and predict the consequences they have on other related natural phenomena. Con-

---

[1] See also John Paul II 1988.
[2] Auletta 2011a, Sec. 16.2.
[3] Margenau 1950, pp. 15 and 167-171; See also Ayala 2007, pp. 184-191.

sequently, we dissociate from a rather descriptive understanding of the function of theories as expressed by some proponents of the neo-positivistic school[4], which dominated epistemology in the first half of the previous century and already characterized the first stage of Western science, as said in Sec. 1.2.

Thus, theories are usually built and elaborated on in order to provide *explanations* of vast ensembles of natural phenomena or of a definite domain of reality. They are generally composed of laws and principles, entities, and explanatory mechanisms:

- *Laws* are the core of the theory and enable it to make predictions about the concerned domain of reality. They establish (formal) relations between initial and final conditions in certain key or idealized processes, i.e. they describe the variation in time of relevant parameters, and by virtue of this they are able to provide predictions. Laws *always* have a deterministic character; however, this does not imply (as it was traditionally assumed) that nature is deterministic in itself or at all levels. Indeed, as we shall see in the following, science is more and more addressing laws that deterministically describe the time evolution of *probability distributions* of certain happenings and/or of *classes* of systems. Therefore, law-like scientific explanations should not always be meant as possessing a causal character if causality is regarded as being exclusively efficient or mechanistic[5]. *Principles* still conserve the formal character of laws but are *foundational* assumptions of a rather general scope and often of a heuristic nature. For instance, the so-called "correspondence principle"[6] is indispensable in order to define observables like position and momentum in quantum mechanics[7]. Other examples are provided by the conservation principles in physics or by the central dogma of molecular biology[8]. It is worth mentioning that principles do not usually constitute the *a-priori* starting point like kinds of premises of our scientific researches; they are rather statements *gained* after thorough investigation, which are then able to guide further scientific research in the concerned field. The importance of this issue will be progressively unveiled in the next three subsections.

---

[4] See Carnap (1928) and Nagel (1961, Ch. 5).

[5] Auletta *et al.* 2009a, especially Chs. 9 and 16. See, also, Auletta 2011a, especially Chs 2, 6, 8, and 24-25; Auletta 2008; Auletta *et al.* 2008; Colagè 2008.

[6] Bohr 1920.

[7] Auletta *et al.* 2009a, Sec. 2.2.

[8] Crick 1958; 1970.

- *Entities* are the actual referents of the scientific enterprise, i.e. what a theory deals with, even if it is not always possible to experience them directly[9]. For instance, at the beginning of the 19th century, J. Dalton[10] assumed the existence of atoms to explain the conservation of mass in chemical reactions and the stable proportions of the involved elements although he had no idea of what kind of characters these atoms could have (this is an understanding that came much later, at the beginning of the 20th century, thanks to the theoretical work of Rutherford, Bohr and others). Similarly, G. J. Mendel assumed his "factors" of heredity, later named "genes", as a plausible explanation of the results of the experiments he performed on plants, and conceived them as the minimal entities responsible for *discontinuous* and recurrent heredity, without having at that time any means of directly ascertaining the very existence and characters of such units of inheritance. Mendel himself thought that the unexpected regularities that he found were only present in some species and only determined some characters. The successive work carried out by scholars like De Vries, Bateson and Morgan eventually provided the experimental and conceptual framework for a generalized model of heredity based on the regularities grasped by Mendel; finally, with the discovery of the genetic code[11] a full theoretical justification of Mendel's insight could be found. Summarizing, entities often have a putative character and are not completely dependent on descriptions due to the incompleteness, the lack of or even the falsity of many explanations at hand[12].

- Finally, according to a modern point of view, scientific theories (at least the non-descriptive ones) need to provide explanatory *mechanisms* that are able to point out precisely what the factors are (which we usually call causes) that trigger a certain process and the specific modalities in which such a process occurs, thus producing certain effects that we consider to be noticeable or otherwise unexplainable. For instance, it has been assumed for a long time that information acquired by the peripheral sensory system is transmitted both in and throughout the brain. The discoveries about the anatomy of the neuron, the specialized molecules called neuro-transmitters responsible for synaptic transmission and the axonal interconnections among neurons are precisely the body of knowledge that constitutes the explanation of how an external stimulus is acquired in its informational value and processed by the brain.

---

[9] Auletta 2003; 2009c; Colagè 2008.
[10] Dalton 1808.
[11] Watson–Crick 1953.
[12] Harré 1986, pp. 97-107.
[13] Duhem 1906, Ch 1.

The assessment of the existence of natural entities in science, as well as of their established properties and effects, represents a fundamental condition for the understanding of what nature really is. This is the order of consideration that binds the scientific enquiry to an ontological perspective. Such an assessment also turns out to be indispensable for the very practice of scientific research, since it determines the appropriate experimental apparata and methodological procedures for a theory to be both *operative* and *applicable*. Indeed, entities may be considered as the theoretical "crystallization" of the scientific results that we *hypothesize*[13], put at work, and progressively refine according to our experimental enquiries and results, especially taking into account the fact that explanatory mechanisms will in general let us discover further and often unpredicted entities. For instance, once von Helmholtz had experimentally excluded the old Cartesian explanation of a pure mechanical transmission of an electrical stimulus to the brain and Cajal and Golgi had described the anatomy of the neuron, the discovery of neurotransmitters was a consequence of the search for mechanisms able to explain how neurons pass information to each other. In our view, then, the undertaking of a mature scientific enterprise cannot be disjoined from the ontological quest. In this sense, modern science not only consists in a search for pure interdependencies or correlations[14] but – in enquiring about the mechanisms that are able to explain certain noteworthy phenomena – it also asks about the factors (causes), and therefore the kinds of entities that may have produced them.

2.1.2 *Foundations, Positivism and Under-determination of Scientific Theories*

The attempt at giving ultimate theoretical foundations to the explanatory power of science is well known. The first modern effort in this direction can be traced back to Descartes' search for the ultimate ground upon which our knowledge must be secured, even in order to progress from its very first steps[15]. This aspiration was also shared by other modern philosophers and scientists. For instance, the idea of a *mathesis unviversalis* that could give rise to a system in which knowledge as a whole (and not only the formal disciplines) were set in a *calculus*, was central in Leibniz's work[16].

---

[14] As still maintained in the neo-positivistic schools: see Carnap 1928, § 165. See also Margenau 1950, p. 28.

[15] Descartes 1641, *Meditatio Prima*; 1644, I, 7.

[16] See the first part of Leibniz *PS*, v. 7.

Immanuel Kant also tried to discover the foundations of human knowledge by attempting to ground the empirical (Newtonian) science of his times. Indeed, the first aim of the *Critique of Pure Reason* is to establish the *a-priori* conditions required for objects to be given to human knowledge[17]. In so doing, he also established clear limits to the scope of human knowledge. Namely, and with a certain consonance with the modern criticism about the essences (Sec. 1.2), he famously affirmed that the objectivity of human knowledge is concerned with phenomena and thus cannot grasp the "things-in-themselves" (the noumena) in any way[18].

The hope of finding complete foundations for knowledge, at least of the formal sciences, in the second half of the 19th century led to the so-called Hilbert's program[19]. This perspective, also known as the *formalist* program, aimed at finding a *finite set of axioms* from which the whole of mathematics may follow and at proving that such a set of axioms is *consistent*. The idea of reducing mathematics to a finite formal system is also present in the works of Frege[20]. In spite of important differences in the approaches of the two 19th-century scholars, their research programs could be understood to a certain extent as a further development of Leibniz's program to build a *mathesis universalis*.

A serious flaw in the project of settling the foundations even of a sole arithmetic was found by Bertrand Russell who discovered a contradiction in the general use of the concept of "class" in the first volume of Frege's *Foundations of Arithmetic*[21]. The foundationalist approach in mathematics was later blocked by the so-called Gödel's theorem[22]. The theorem, indeed, *demonstrates* that any non-contradictory formal system at least so powerful as to include arithmetic, always contains a proposition that, although taken to be true, cannot be proved by its axioms. If such a result is true of formal systems, *a fortiori* the task of a deductive foundation of human knowledge as a whole appears to be fully hopeless.

---

[17] Kant 1787, A16-B30.
[18] Kant 1787, Introduction to the second edition (1787).
[19] Hilbert 1903; 1915
[20] Frege 1884.
[21] Russell 1902; 1903. The error that Russell found is concerned with the fact that both the definitions of "the class of the classes that are not members of themselves" and of "the class of the classes that are members of themselves" (and similar classes) lead to a contradiction that cannot be resolved within the system that defines the class itself.
[22] Gödel 1931.

From Descartes to Hilbert, foundationalism, despite the different forms it has taken, has had a clear task: to warrant human knowledge, or at least some of its domains, as deriving from a-priori principles. In our understanding, foundationalism has actually missed the essential character of the cognitive enterprise, which consists in a programmatic research, in a *projection*, through which, by formulating hypotheses of increasing generality, we reach, in a progressive manner, more solid foundations[23]. Therefore, the kind of necessity that science – and cognition in general – can find is not identified in the presuppositions as such but in the ability to progressively connect different theoretical aspects and gathered data with pondered guesses in a more concise and productive way. The assumptions that become basic principles for science are therefore those that have been critically examined in the light of the subsequent enquiry[24]. Therefore, knowledge proceeds from merely probable antecedents to the sufficient proof of a fact or a truth, and, after the proof, to the acquisition of a certitude about it[25]. As J. Haughey eloquently asserts, we anticipate wholes in a succession of tentative solutions instead of being determined by them[26].

The crisis of foundationalism led to the conviction that there cannot be a unique scientific explanation of reality and that science is made up of a plurality of approaches and disciplines whose points of view do not necessarily fit one another. The failure of this program also supported the epistemological thesis of the underdetermination of scientific theories, according to which no theory can fully account for a certain domain of reality[27]. As a consequence, the possibility that

---

[23] As we shall see, this is an Aristotelian point of view: see for instance the basic distinction in *An. Post.* (71a25-30) about to know *universally* in the sense of a knowledge that is not fully specified or determined and to know *simpliciter* (*haplos*); for the last term see Barnes' edition and commentary. Moreover, Wedin (2009, p. 135) says that the principle of non-contradiction (the ultimate but formal principle, according to Aristotle) "is the doctrine that everyone who demonstrates goes back to in the end – not as the principle from which all deductions start". Einstein (1936, p. 96) said that physics is a theory in evolution "going in the direction of increasing simplicity of the logical bases", a process that deals with the justification of the system, provided by the proofs and generalizations of the theorems we formulate on the basis of the experience.

[24] Maréchal 1926, v. V, pp. 217-236.

[25] Newman 1870, Ch. 8, § 2.

[26] Haughey 2009, p. 57.

[27] Therefore, underdetermination is concerned with the fact that no explanation can give an exhaustive account for a body of data and *not* with the fact that there are actually alternative

sooner or later another theory will be found that solves the same kinds of problems in a better way can never be excluded. This can also be seen from a historical point of view by looking, for example, at the replacement of Newton's gravitational law by Einstein's general relativity.

A specific and interesting form of the under-determination of theories is the so-called "Duhem-Quine thesis"[28], stating that a single and *isolated* theory or hypothesis cannot be checked as such as it always needs other items of knowledge (back-ground assumptions and/or auxiliary hypotheses) for drawing predictions to be compared with data. Therefore, it is never a single hypothesis but a *bundle* of hypotheses and theories to be checked. If, moreover, such a bundle of theoretical assumptions is falsified by certain evidence, then it is not possible to establish with certainty which single hypothesis in the bundle is the one to be actually false, since the only way to do this is *to prove* the truth of *all* the other hypotheses. The latter task, however, cannot be achieved entirely; therefore, not only are we unable to reach certainty about the truth of a theory, but according to the Duhem-Quine thesis, neither are we able to conclusively and definitely falsify a theoretical corpus.

This also means that we cannot approach empirical data and evidence without the presupposition of a certain conceptual framework, thus implying that empirical evidence is always interpreted and organized according to theoretical commitments. It is worth mentioning that, as to this, the positions of Duhem and Quine are not exactly coincident. In Duhem's view such commitments concern specific scientific fields like *physics*. Moreover, the "bundle of theories" includes physical propositions and metaphysical assumptions only[29]. Along this line of approach there also seems to be a recently proposed model-dependent realism, according to which there is no picture- or theory-independent concept of reality[30]. Since, according to this view, it becomes pointless to ask whether a certain model is real but only whether it agrees with observations, we could not say whether the Ptolemaic model of the universe is more real than the Copernican one or vice versa. Quine's position, however, is even more ex-

---

theories, as Harré (1986, pp. 74-75) seems to assume. As a matter of fact, in the historical development of science, in most cases there are no alternative theories for the same set of data. The true logical consequence is that, since theories are underdetermined, then, alternative explanations are *possibile*.

[28] Duhem 1906. Quine 1948 and 1951.

[29] A similar position has been supported by Margenau (1950, pp. 12-16 and Ch 5), and, as we shall see in the following, it is somehow more compatible with our overall perspective.

[30] Hawking-Mlodinow 2010, pp. 39-59.

treme. Indeed, in his view, the "bundle of theories" ultimately covers the *whole of human knowledge* in a particular historical epoch (and in a particular geographical area): as a consequence, the under-determination of theories implied by the "Duhem-Quine thesis" seems to apply, for Quine, to human *culture*[31]. From this radical position, it seems that the logician has drawn the conclusion that human knowledge is a pure social construct[32].

In contrast to foundationalism, we have the idea, rooted in Empiricism, that scientific knowledge is based solely on pure phenomenological sensorial experience. This position has been systematically proposed by Ernst Mach[33] and, later by R. Carnap[34]. The core of Mach's criticism of Boltzmann's atomistic theory is well-known: since at that time it was impossible to have any (even indirect) sensorial evidence of atoms, they should be considered as purely fictional entities without any ontological import and therefore, from Mach's perspective, also deprived of scientific value. Indeed, even lacking sensorial experience of atoms, Boltzmann identified them as the ontological referents responsible for some observed phenomena especially related to the behaviour of gases. However, at the beginning of the 20th century atoms were shown to be real entities, or at least to have an ontological import. Therefore, postulated or inferred entities, which cannot be well described or experienced at a certain moment, can turn out to be an object of (at least indirect) experience later on. It seems that here Mach misses the crucial distinction between fictional entities (like Democritus' atoms) and theoretical entities (like Dalton's and Boltzman's atoms) that are introduced or postulated in the context of a certain scientific theory as explanatory factors to foster further research. Indeed, hypothesizing the existence of these entities often leads to the consideration of which particular consequences they imply, thus helping science to overcome precisely those limitations that Mach or neo-positivists considered as inviolable. The above misunderstanding stems from Mach's assumption of a sharp separation between what is verifiable and what is purely fictional, and therefore not verifiable (somehow following here the mentioned Kantian distinction be-

---

[31] Quine 1960.

[32] Quine 1969. See also Auletta 2003.

[33] Mach 1905. For the empiricist tradition see Hume 1748. We also acknowledge that Mach's position is more complex since he could be considered as the beginner of a tradition that has given rise to the *Gestalt* as well (see Fabro 1941a, pp. 81-82).

[34] Carnap 1928. His rejection of any referential relation as being constitutive of science (§ 15) is especially enlightening.

tween phenomenon and thing-in-itself). We may suggest that empiricist and phenomenological epistemologies only account for a first and often immature stage in the building of scientific theories, namely the search for some empirical regularities. However, as explained (Subsec. 2.1.1), a scientific theory must also provide models and general explanatory mechanisms for the problems at hand. As noticed by R. Bhaskar[35], a scientific theory is fully accomplished by a third step, that is, the ontological inference about the possible entities responsible for those mechanisms leading eventually to acknowledge the existence of new realities.

Consequently, the later neo-positivistic epistemology of the theories' verification[36] (derived from Mach's own positions) did not seem to be adequate to account for the real work of science. These epistemologies pertain to the category of so-called truth-realism, which assumes that theories are pure descriptions of reality and that realism is founded on the notion of truth as a correspondence with empirical data[37]. If Gödel's theorem blocked the possibility of founding formal constructs "from above" (i.e. by deriving them from a certain set of warranted or self-evident axioms), it is also true that it is not possible to confirm a theory *definitely* on a pure empirical basis either[38], as the history of science shows.

Summing up,

- One cannot settle the absolute a-priori foundations of scientific research,
- Science cannot be reduced to sensorial experience because of its wide-ranging ontological commitment,
- Scientific theories cannot be fully verified.

In conclusion, we may say that both foundationalism and neo-positivism have not only ill posed the problem of the foundations but also missed the issue of the *ontological import* of scientific theories. Our standpoint is that theories are able to recover and reinforce their connection with reality *in a dynamic process of continuous adjustment and self-correction of certain hypotheses.* In other words, the building of scientific theories can be led back to the procedures that science employs in embed-

---

[35] Bhaskar 1975.
[36] Carnap 1932; Neurath 1932-1933.
[37] Harré 1986, pp. 65-67.
[38] Popper 1934.

ding new experiences within a dynamic theoretical framework continuously tested, enlarged and corrected[39]. Such procedures are essentially inferential and applied to empirical results.

In the following, we shall try to show that a renewed view of inferences may cast some light on the issues of the foundations of the scientific enterprise and of its ontological import[40]. We aim at showing that science makes use of three fundamental forms of inference, namely: abduction, induction, and deduction[41]. We would like to start with deduction and then focus on abduction as it often constitutes the first inference coming into play when scientists are faced with *problems* in their theories, and therefore most appropriate when enlightening the dynamical nature of scientific theories. In the next chapter, we shall come back to induction.

2.1.3 *Science is Based on Inferences*

As explained in the two previous subsections, theories (especially postulated entities and their relations with regularities expressed in laws) must first be formulated in a *putative* way and then tested, revised, *corrected* and *"mopped up"*[42] in a historical process sometimes taking centuries (it suffices to recall the vicissitudes of the so called gravitational "force" throughout the history of physics, Sec. 1.3). We have indeed just shown how difficult it is to conceive of scientific theories as being built by starting *only* from some set of already ascertained items of knowledge[43]. At any given moment, the "pure" facts ascertained beyond any possible doubt and the punctual explanations for specific phenomena and empirical occur-

---

[39] Poincaré 1902, pp. 157-161. See also Margenau 1950, p. 105.

[40] Bhaskar 1975.

[41] Auletta 2009a.

[42] Kuhn 1961.

[43] It is interesting to remark that A. Comte, the father of positivism and often erroneously considered as the grandfather of the cult of facts characterizing neopositivism, says (1830-1842, II, Lect. XIX) that astronomy shows a superior intellectual and rational dignity relative to other sciences precisely because it is based on inferences. We also agree with his statement that what is distinctive about natural sciences is their ability to predict certain consequences. Obviously, we disagree about his sharp separation between science and what he calls metaphysics and theology. As said, we consider the traditional associationist philosophy characterizing Hume and especially J. S. Mill (1865, pp. 9-10) as largely insufficient to account for the dynamics of knowledge; see also Fabro 1941a, Chs. 1-2 and Auletta 2011a, Sec. 12.2.

rences are extremely limited. Scientific research unavoidably proceeds by means of hypotheses that are based on the few acknowledged facts available and that are dynamically put at work in a canalizing historical process toward better and better (but never completely exhaustive) explanation of some piece of reality. Therefore, the ultimate criterion to judge scientific hypotheses should not be the mere correspondence to facts, but their fecundity and prospective value for the progress of knowledge itself (i.e their non-*ad hoc* character). Sometimes, this fecundity may become apparent in the course of decades.

These preliminary considerations already allow for the understanding of the relevance of *error correction* in science[44]. As a matter of fact, it is often the case that hypotheses acknowledged to be correct are supported by false "proofs" and it is only later on that correct evidence is found. Let us give a well-known example: Galilei took the tides induced by the moon as evidence for the correct hypothesis of the Earth's motion[45].

Notwithstanding its putative origin, it is crucial to a scientific theory (or even to a narrower hypothesis) to be *testable at least in principle* or to have empirical consequences. To formulate predictions or to consider what the possible consequences of certain theoretical assumptions could be is the business of the form of inference called *deduction*[46]: once a theory is established, or even when certain hypotheses are clearly formulated, the scientific activity is mainly concerned with the ways in which these assumptions could affect our understanding of nature, as well

---

[44] "And what do we mean by the real? It is a conception which we must have had when we discovered that there was an unreal, an illusion; that is, when we first corrected ourselves." (Peirce 1868, p. 239). See also Peirce 1868, p. 212. Error is not simply ignorance or illusion (Hamilton 1859, IV, Lect. XXVIII, pp. 78-80) but is determined by some form of conflict either between different domains of what is known or between what is known and what is still unknown or not very well known.

[45] Galilei 1632, Giornata quarta.

[46] It is clear that here and in the following we are speaking of natural deductions or of science-guided deductions and not of formal ones (Aristotle in the Ch. 2 of book I of *An. Post.* distinguishes between deductions and demonstrations, an opposition that could be understood in terms of the distinction natural deductions/formal deductions). Natural deductions do not receive sanction through mechanical or computer-like calculation but through possible interesting consequences or also what along with J. H. Newman (1870, Ch. 9) can be called *illative sense*, i.e. the power common to all humans to control the operations of one's own mind.

as how they could dismiss previous assumptions or affect related fields of investigation. Therefore, deduction is responsible for singling out the *expectations* of our scientific assumptions and hypotheses. The very deductive derivation of predictions and empirical consequences gives to scientific theories the possibility of being compared with data and evidence.

The comparison between expectations and evidences may back-affect some previous assumptions and hypotheses, so that science turns out to be intrinsically characterized by a continuous process of pondered trial and error, by means of which it is able to progress, correct errors and *progressively found* itself. So, the driving force of knowledge is represented by problems that raise challenges that we formulate in questions; from questions we go on to hypotheses in order to solve them and to correct errors. Scientific progress and error correction are primarily a matter of the inferential reasoning called *abduction*[47]. Abduction is the kind of inference that leads to introduce a *new property* or behaviour into a certain scientific investigation. Given a set of entities that have a certain character in common, abduction splits such a sample into two subsets, one of them collecting the entities that share an already known property or behaviour (therefore successfully framed in a previous theory), and the other gathering entities that are assumed to be characterized by a new property, thus representing a new class of objects. An abductive inference is generally triggered when we find a surprising class of cases such that, when we apply the established laws (or theories), the consequences turn out to be unpredicted and *seem* even to contradict them. For instance, Planck found that the classical electromagnetic theory, *when applied* to a black-body (i.e. a hollow sphere with trapped radiation inside), showed discontinuous properties that seemed to contradict the classical treatment of light as undulatory. However, according to Planck[48], light *in itself* still followed the classical theory, and only when reflected by the internal surface of the black-body, which is *discontinuous* (being constituted by molecules), acquires discontinuous features (this is precisely the *new* property).

In general, therefore, abduction is the inferential process that comes into play when surprising and unexpected facts force us to correct our approach *not* by

---

[47] Peirce 1866; 1877; 1878a. See also Auletta 2009a; 2011d. Popper's (1934) falsification is an epistemological development of Peirce's fallibilism that we shall consider in the next subsection.
[48] Planck 1900a; 1900b.

changing the theory but by modifying its *application* to a new, unpredicted class of objects (eventually defined by a new property). Here, it is worth stressing that, being an inference, abduction should not be confused with the sudden and positive grasping of a new property applicable to the abductively inferred set of objects, an act that we call *insight*[49]. Abduction is rather a *prerequisite* for such an act, showing the necessity of finding a *positive* characterization of this new set of entities initially singled out by negative discrimination (these objects do not satisfy such and such a property…). Therefore, insight is the natural but often genial *fulfilment* of the inquiry started with the identification of a problem and developed by means of an abductive inference. Given the hypothetical character of abduction, there is no certainty that the real entities are exactly as the theory supposes them to be; nevertheless, without such reference to entities the progress of science would be significantly hampered.

2.1.4 *Realism and Fallibilism in Science*

Now, if abduction plays such an essential role in scientific inferential reasoning, what does ensure that science itself actually informs us about reality? After all, it might be a mere product of imagination, thus the entities it supposes would be merely fictional. At first, the answer to this question is the simple acknowledgement that scientific theories work very well throughout natural reality; they allow for careful predictions about nature and successful interactions with it, even when they postulate theoretical entities (Subsec. 2.1.2). As stressed by H. Putnam[50], realism is the only account of the way scientific theories work without making the success of science a miracle. Therefore, it is at least reasonable to think that science is able *somehow* to grasp reality as it actually is, and when we have an experience that seems to us to be conclusive for a certain enquiry, we express this in the form of an assent.

---

[49] Lonergan 1957, p. 28. Einstein said: "physics constitutes a logical system of thought which is in a state of evolution and whose basis cannot be obtained through distillation by any inductive method from the experience lived through, but which can only be attained by free invention" (1936, p. 96). See also Auletta 2011a, Subsec. 18.4.3 and Sec 21.3. It is well known that Plato developed a theory of insight, as it is witnessed e.g. by the so-called 7th Letter (341c-d). It is also possible that the modern doctrine of the intellectual insight has a root in Avicenna's theory of illumination (GR, II, Secs.VIII-IX; III, Secs. VII-VIII and XIX).
[50] Putnam 1982.

It is true that, during the history of science, there have been several attempts at reducing scientific enterprise to a pure mathematical formalism. We recall here the example of Heisenberg[51] claiming that it is only mathematical formalism that really matters in quantum mechanics, mainly due to the difficulty in *observing* quantum-mechanical entities. The actual relevance of formal constructs in science, as well as their ontological status, will be one of our main philosophical concerns in the following study. A related problem is a consequence of Quine's radicalization of the Duhem-Quine thesis (Subsec. 2.1.2). Such a position is the so-called "*referential opacity*", proposed by Quine himself[52] as a consequence of the load of interpretation on any referential act: it states that, when using a term, it is not evident if it *refers* to a whole entity, rather than to some of its parts or even to some of its configurations. This idea has an actual relevance in linguistic contexts, especially when translation is concerned, and at a representational level in psychology. However, it seems that, in Quine's view, this implies an "ontological relativity"[53]. In fact, recent studies in psychology of development have shown the unfoundedness of the thesis that there is such a general opacity of reference. Indeed, children are able to solve the problem of reference thanks to a set of implicit assumptions, the most important of which is that a term is taken to refer to a whole object, and not to its properties or parts if there are no additional assumptions[54].

After our previous critical examination, we would like to positively summarize in a few words one of the two columns of our epistemological approach (the other one being fallibilism), namely realism: actual breakthroughs in science (like the understanding of the hydrogen atom, the discovery of non-local correlations in quantum mechanics or of the DNA structure in biology) have been achieved by asking what kind of entity might be responsible for some observed and unexpected behaviours or anomalies. Without such kind of ontological questions, the whole scientific enterprise would be reduced to a verbal game. Scientists assume their putative theoretical objects to be good descriptions of real existing entities, even when they are aware that the descriptions associated with those objects may well turn out to be wrong in the light of further theoretical and experimental research. The

---

[51] Heisenberg 1927. See also Auletta 2000.

[52] See Quine 1951; 1960.

[53] Quine 1969, pp 28-68.

[54] For more details, see Auletta 2003; Auletta 2011a, Sec. 20.6. See also Markman 1990 and Markman–Wachtel 1998.

40

very hypothetical assumption of some entities represents a provisional attempt at *intelligibly understanding* the concerned portion of natural reality. Subsequently, provided that the research itself has furnished the means to support these assumptions, entities will become a well-established piece of knowledge upon which natural science may rely for the furthering of its own activity. Consequently, the progress of knowledge can be conceived of as reflecting three fundamental kinds of mental acts: doubt, inference, and assent. A question is the expression of a doubt (the manifestation of a problem); a conclusion is the expression of an act of inference; and an assertion is the expression of an act of assent[55].

All the instrumentalist and anti-realistic conceptions of the scientific enterprise consider scientific theories as being in principle both interchangeable and arbitrary constructs that are devoid of any ontological import. This is not deprived of a certain ground. As we have already mentioned, the main reasons for such opinions are the acknowledged failure of the foundationalist approach (Subsec. 2.1.2) and the history of science, which shows many amendments and turns. We have already mentioned that in some cases theories previously maintained to be true have been eventually proved to be false in spite of the fact that they previously worked. It also happens that theories that, up to a certain degree of exactitude, still work quite well within a certain scope, although they cannot be successfully applied outside[56], which is the case with classical mechanics. These facts rule out the possibility of opposing to the anti-realistic views of science by way of *naïve* realism, that is, any position maintaining that scientific theories *describe* reality as it actually is in a faithfully *mirroring* way. It is therefore necessary to develop a view of science that puts together both *realism* and *fallibilism*[57].

Peirce defines *fallibilism* as the doctrine maintaining that it is impossible, for human knowledge, to attain absolute certainty, absolute exactitude and absolute universality[58]. Indeed, also very basic and well-established principles and assump-

---

[55] Newman 1870, Ch. 1, § 1. On the notion of assent see also Locke 1689, B. IV, Ch. 15. For the relevance of questioning see also Lonergan 1963. We thank Cardinal C. Ruini for having attracted our attention on the last paper.

[56] Auletta 2007a; Colagè 2006.

[57] Peirce *CP*, 1.149-151, 1.162, 2.227; Peirce 1903b, p. 227; Popper 1963; Auletta–Tarozzi 2004a; 2004b.

[58] Peirce *CP*, 1.108-120.

tions can be put into question by further empirical achievements. For these reasons, the approach known as Poincaré's conventionalism turns out to be at the very least inaccurate. In short, Poincaré assumed that the principles of classical mechanics (the dominant science at that time) are definitions that can never be contradicted by experience[59]. In fact, quantum mechanics shows that many general principles assumed by classical mechanics, like the principle of least-action, are no longer applicable (at least in their original formulation) to other domains of physical reality[60]. This implies, however, neither a desperate pessimistic view of science, nor an anti-realistic attitude. Indeed, according to Peirce's fallibilism, scientific enterprise, although subject to errors, is also intrinsically characterized by a continuous process of *self-correction*[61] that enables us, over a very long period of time, to *asymptotically converge* to truth, or at least to have a growing probability of being correct in our inferences[62]. This is the reason why Peirce, criticizing Berkeley's idealism, affirmed: "Where is the real, the thing independent of how we think it, to be found? There must be such a thing, for we find our opinion *constrained*; there is something, therefore, which influences our thoughts and is not created by them"[63]. The first patent consequence of such a process of convergence is that science's predictive power, empirical adequacy, and applicative reliability all increase with time. More importantly, such an increase allows in turn for *inferring* more grounded *interpretations* about the fundamental characters and constitution of nature.

This enables us to distinguish between scientific enterprise, which is characterized by the ability to formulate correct predictions, and the cognitive journey (also involving philosophy and theology) that finally aims at truth (that is, the full and total understanding of reality). In other words, truth is not the criterion with which we judge theories (which only have a partial and fallible understanding of reality), but we are confident that our growing body of scientific knowledge allowing us to

---

[59] Poincaré 1902, pp. 123-124.

[60] Margenau 1950, pp. 76-77.

[61] Peirce 1898b, p. 44. See also Poincaré 1902, pp. 165-166 and Lonergan 1957, p. 372.

[62] Peirce 1869, pp. 268-269; 1871, pp. 468-469; 1878b, p. 273. For this reason, although a tolerance of the other's opinion is very important (and culturally relevant) and often put in danger by all kinds of fanaticism (Draper 1875, Ch. 8), we do not need to elevate the diversity of opinion to a final goal in itself.

[63] Peirce 1871, pp. 468. See Sellars 1920, pp. 3-6.

make finer and finer predictions will also lead us to truth[64], especially when the former is appropriately integrated with philosophical and theological insights.

It is precisely here that we find one of the most relevant limits of the Kantian distinction between noumenon and phenomenon (Subsec. 2.1.2). Any time we successfully correct our theories and hypotheses, we unveil *some* real feature of nature as it actually is: it is indeed nature itself that forces us to correct our assumptions, otherwise the necessity of a correction would not have arisen at all. As R. W. Sellars puts it[65], the most important contribution that Idealism brought to philosophy is the emphasis on the activity of the knowing-subject; Idealism's error, however, is the conviction that such activity makes it immaterial to reach reality in its independent existence (and therefore also the truth about it). Indeed, Kant's philosophy implies that the cognitive structure of the human subject is totally unconstrained by reality[66], since any constraint is again due to the imposition of that structure on experience, in a way that somehow recalls Quine's referential opacity. Obviously, it remains true that we cannot have a direct experience of an independent reality. However, an opportune combination of experience and inference can let us reach interesting ontological conclusions. Appropriately, R. Bhaskar has called "epistemic fallacy" the reduction of statements about being to statements about knowledge, or ontological questions to epistemological ones[67]. The crucial point is that the constraints that Peirce thought that reality imposes on our beliefs are not there from the start (as many foundationalists would claim), nor will they be imposed ultimately in a hypothetical "final" knowledge. On the contrary, nature exerts them *during the course* of our scientific cognitive enterprise. In this course, we are able to build an ontology that is not constructed a priori, but is nevertheless not a mere consequence of our hypotheses.

---

[64] See Benedict XVI (2011). It seems to us that Harré does not fully distinguish between these two aspects (Harré 1986, pp. 41-42).

[65] Sellars 1920, pp. 143-150.

[66] Lonergan (1963) quotes the work of Emmerich Coreth, *Metaphysics* (Vienna, 1961), who says that from the fundamental questioning attitude of humans it follows that there is no closed inner space of a trascendental subjectivity, since subjectivity in its questioning is simulateneously outside (*draußen*).

[67] Bhaskar 1975, p. 36.

2.1.5 *Critical Realism*

The interpolation of realism and fallibilism leads us to critical realism. The historical root of critical realism can be found in a paper of William James on Spencer[68], while the term "critical realism" was introduced for the first time in Roy Sellars' works[69]; Ian Barbour reignited the debate about this issue in a theological context in the sixties[70]. Another relevant source is represented by the contribution of B. Lonergan[71]. The kind of critical realism we are envisaging here is strictly connected to what R. Harré called "referential realism", due to his stress on the referential import of scientific theories in their operative interaction with reality[72]. The same holds true for Bhaskar's transcendental realism, considering in particular its stress on ontological inferences[73].

A starting point towards a critical realism is the consideration, drawn as a conclusion of our previous examination (Subsecs. 2.1.2-2.1.4), that theoretical constructs are not exhaustive descriptions of nature. According to critical realism, on the contrary, scientific theories *unveil* reality but cannot and even do not need to build a "pocket-world" as an isomorphic model of it. In the words of J. Polkinghorne, "scientists are mapmakers of the physical world. No map tells us all that could be conceivably told"[74]. Also Sellars stressed the point, at the very beginning of his *Critical Realism*, by saying that the "physical" world cannot be *like* our ideas, thus encouraging us to relinquish all attempts at *picturing* reality. Instead, he claimed science to be *tested* knowledge about the world[75], and therefore deeply embedded in a dynamic process of growing understanding. It can also be added that even our perceptions do not mirror the properties of the objects that we experience. However, it is not true that our perceptions are only ideas

---

[68] James 1878.

[69] Sellars 1920; 1922.

[70] Barbour 1966, pp. 284-286. For an examination of the issue of critical realism see also McGrath 2001-2003, v.2, Ch. 10, especially pages 195-226.

[71] Lonergan 1957, pp. 357 and 366. See also Polkinghorne 2009, p. 25.

[72] Harré 1986, pp. 8-95.

[73] Bhaskar 1975.

[74] Polkinghorne-Beale 2009, 84-85. See also Peirce 1869, p. 249.

[75] Sellars 1920, p. vi.

[76] In philosophy, it is difficult to find the first position about perception in general, but many philosophers who have supported the distinction between primary and secondary qualities assumed that the ideas of the former are resemblances of them (Locke 1689, II, Ch. 8, § 15). In general, most modern philosophers share the idea that truth consists in some form of re-

in the mind either[76]. These two opposite views seem to miss the essential fact that our perceptions are evoked and articulated responses to certain stimuli that we receive[77]. Therefore, we know things only through their effects[78] and our perceptions are perfectly objective although fully relational and interactional, a point that has been very often misunderstood[79].

Polkinghorne considers critical realism to be a very useful position, especially in the dialogue between science and theology, and points out six fundamental characteristics of it[80]: (1) piecemeal achievements in knowledge instead of total accounts, (2) resistance to any attempt at defining, or distilling the essence of science, (3) the existence of a self-sustaining circularity in the mutual relationship of theory and experiment, (4) no universal epistemology since entities are knowable only through methods that conform to their idiosyncratic nature, (5) social factors as not determining scientific research, although playing a role in accelerating or inhibiting its growth, and (6) the doctrine of scientific realism as the best means to understand the actual experience of doing science.

Point (3) of Polkinghorne's presentation of critical realism seems to us to be able to clarify the specific way in which this doctrine maintains that science actually grasps reality in an operational and interactive way[81]. Some might understand crit-

---

semblance of our representations with external objects, and this tradition is still at the root of Kant's ditinction between noumena and phenomena. The second philosophical position has been supported by several philosophers, like Malebranche (1674-1678) or Berkely (1710) and to a certain extent by German Idealists.

[77] This point has been very well understood by Reid (1764, Ch. 6, Secs. 5-6). It is worth recalling here that J. Locke (one of the targets of Reid's criticism as far as he opened the path to idealistic positions: Reid 1764, Ch. 6, Sec. 4 and Ch. 7) had a position that was more complex (as the previous footnote also shows). Indeed, Locke defines (1689, I, Ch. 1, § 8) an idea as any object of human understanding but also adds (Locke 1689, II, Ch. 1, § 3) that ideas, at a primary level, are formed in the mind when objects affect our senses and stresses that the primary qualities of bodies are powers that produce various sensations in us (1689, II, Ch. 8, § 10; II, Ch. 23, § 28). As is well known, Locke even considers particles of matter as being the final source of this action on our senses (1689, II, Ch. 8, § 9) so that he finds in a corpuscularian hypothesis an intelligible explication of the qualities of bodies (1689, IV, Ch. 3, § 16). This point has been criticized in Auletta 2002.

[78] Hamilton 1859, I, Lect. VIII, pp. 137-141.

[79] Auletta 2011a, Subsecs. 1.3.3, 2.3.1, and 4.4.5.

[80] Polkinghorne 1998, pp. 105-109. See Torrance 1985.

[81] Colagè 2010.

ical realism to be a non-realistic theory as such. It is clear that critical realism implies a view of science in which the subject is deeply involved (both in formulating hypotheses and testing them). However, we can ensure a partial linkage between our theoretical constructs and empirical facts precisely thanks to the emphasis on testing and experiments: it is here that the two poles represented by our hypotheses and reality meet and produce an insightful spark[82]. Therefore, it is in this process that we actually unveil reality not only in the case in which our theories work but also when they do *not* work[83]. Indeed, even false theories may produce interesting results; the reasons why a theory is accepted or preferred to another one is only because it works better and not because one is proven to be true and the other to be false. Obviously, this does not diminish the fact that the results that we find in this way may have ontological import and therefore lead us to enlighten domains of reality.

This point is also strictly connected with our view of the inferential character of the scientific enterprise. By means of abductive inferences and positive acts of insights (Subsec. 2.1.3), scientists hypothesize the constitution of *a certain domain* of natural reality, according to Polkinghorne's point (1). Then, they apply their suppositions to reality by performing well-designed experiments thus *testing* their theoretical elaborations, and, in the case of failure, revise and correct the latter. Therefore, we agree with Polkinghorne's Point (6) that critical realism is a fairly good description of the way in which science is actually practised.

2.1.6 *The Scientific Claim for Universality*

Provided that we are satisfied with this appraisal, it is easy to realize that the question about the *general intelligibility* of nature has only been partially answered, and that the core of the problem is still to be tackled, which will inevitably require further considerations. In fact, we may wonder how far the "game" of scientific disciplines approaching reality in a convergent way could be pushed. To what extent may our scientific theories reach general conclusions? Potentially, there are no a-priori limitations[84], and one expects scientific theories and explanations to be

---

[82] See also Sellars 1920, pp. vi-vii, 38 and 152-153.
[83] See also Farber 2005.
[84] Morowitz 2002, p. 42.

more and more general[85]. This is due to the fact that knowledge ultimately aims at truth (Subsec. 2.1.4), that is, at a comprehensive understanding of the world. As Whitehead[86] said, nothing can be conceived in complete abstraction from the system of the universe. The recurrent philosophical attempt at building a unified science in an encyclopaedic form with very precise requirements about its constitution, scope, nature, and contents[87], is a clear manifestation of such aiming at truth. Also in science there are several unifying (and partly superposed) attempts: (1) The search for a Grand Unified Theory[88] (GUT) among basic forces (electromagnetism, weak and strong forces), (2) the Theory of Everything (also encompassing gravity) or the so-called supersymmetry aiming at showing that matter particles (like electrons and protons) and force particles (like photons or gluons) are two sides of the same coin, but (3) especially the grand theoretical construction known as string theory[89] (which emcompasses results of the latter theories). Such a tendency, moreover, turns out to be heuristically remarkable as it helps to avoid assuming *ad hoc* hypotheses[90], i.e. to prevent scientists to formulate new, particular and "local" hypotheses for any specific problem that a theory may face. As a matter of fact, string theory tries to derive the elementary particles of the so-called standard model (12 carriers of the different forces, 12 leptons, and 36 quarks) from a very small set of assumptions, in particular by postulating that the world at a very basic level is made by (closed or open) *strings* that are attached in different ways to hypersurfaces called *branes* (all the particles are a consequence of their vibrational modes). The theoretical framework is impressive being based on several theoretical columns: Lagrangian mechanics, classical electrodynamics, special and general relativity, quantum mechanics, quantum field theory, and particle physics. Moreover, the connections established among these different fields and the common laws found in this way are astonishing. It is still early to judge what the results will be. However, there is something that is very relevant here for the philosopher and the theologian. We have indeed seen up to now (and shall also see in the following) how important are inferences in science. In general, through inferences one tries to es-

---

[85] Einstein 1918, pp. 108-109.
[86] Whitehead 1929, p. 3.
[87] Wolff 1730, 1731, 1736, 1737; Comte 1830-1842; Neurath 1938.
[88] Barrow 1990.
[89] Zwiebach 2004.
[90] Lakatos 1978.

tablish connections between distant fields or problems or to guess what the solutions of certain problems could be. With the string theory, however, we have for the first time the creation of a huge and systematic theoretical building completely anew and without any empirical support. Will this prove to be a fertile step and with what consequences? The problem is relevant since it still remains true that no scientific enterprise can exist without setting very precise *limitations* of its scope, without individuating the specific (experimental) tools for dealing with its problems, and without careful definition of its particular methodologies. We have indeed remarked that modern science exists by "thinking small" (Sec. 1.2). So, we have here a sort of potential contradiction: science is *universal* in its ambition but *local* (domain-bound) and limited in its status. How can we solve this conflict?

The trend to universality may be expressed in distorted forms when for instance some scientists try to elevate science to a new philosophical and theological (or, equivalently, to an anti-philosophical and anti-theological) global vision of the world, i.e. a *Weltanschauung*. This happens whenever a scientific discipline or even a scientific theory within a discipline is extended far beyond its scope, and the discipline or the theory is eventually considered as providing *the* key conceptual *interpretation* of all dimensions of reality, including the human beings' unceasing quest for ultimate truth or meaning. This is the case, for instance, when evolutionary biology is meant as a cognitive enterprise in competition with theology[91], and/or when religion is considered as an epiphenomenon that could be explained in terms of the biological struggle for survival and therefore also completely overcome, as it happens in the context of some socio-biological approaches[92]. It is well known that R. Dawkins considers science, Darwinism in particular, and religion as *alternative* explanations[93]. Unfortunately, there is a long-lasting tradition that considers science to be an explanation of the world that is alternative to religion and even having the aim of fully removing or annihilating the latter[94].

---

[91] Dawkins 1986; Dennett 1995.

[92] Wilson 1998.

[93] Dawkins 2006, pp. 72-73, 82, and 92-93; see also McGrath 2005, v.1, pp. 51-52.

[94] This was clearly the opinion supported by Comte (1832-1840). Draper (1875, Ch. 9) says that there are two different conceptions of the government of the universe: by Providence and by natural Law. It is however unfortunate when this famous author considers Newton to have shown that the solar system proceeds without any intervention by God, since his theory was quite the opposite as shown in Koyré 1957. On these general matters see White 1896, who

The opposite possibility is to understand science as being solely confined to specific problems. In most universities and academies, research is currently practised as a sort of *applied* science that is surely very useful for technological consequences, but not so insightful when the issue is to find out general *strategies* or ideas that determine the research programs in the long run.

Here, we support the idea that, from the very core of the scientific enterprise, needs for new research strategies and generalizations are continuously rising. For reasons that we have already explained, philosophy and theology can be very helpful to scientific theories for managing such kinds of problems. We would like to consider here two very relevant examples from the history of science of the 20th century. The first one concerns the problem of measurement in quantum mechanics. It is well known that such a problem amazed the community of physicists when von Neumann showed that a measurement outcome does not obey the laws of quantum mechanics and should be considered as a purely random event[95]. A part of the scientific community, including some of the most prominent physicists like Wigner[96], tried to account for this finding by assuming that the human mind had the power to change the state of a quantum system. A philosophical reflection upon this hypothesis showes it to be untenable due to epistemological and ontological difficulties: (1) it provides an explanation of a physical phenomenon without referring to physical mechanisms and (2) it implies a direct influence of the mind on a physical system. With regards to Point (1) we cannot use explanations that consider factors or mechanisms that lie outside the concerned level of investigation. With regards to Point (2), a dynamical action of the mind on physical reality would represent a violation of the causal closure of the physical world (which can also be extended to any level or domain of reality). The second example is provided by the reactions of some cosmologists when the hypothesis of a universe in expansion was first introduced[97]. The latter sounded really extraordinary when one

---

also devotes a chapter (I, Ch. 4) to the antinomy God's intervention/natural Law and even speaks (for instance, I, pp. 22 and 202-208) of the victory of science over theology, as if any progress of knowledge did not always overcome any previous prejudice, also of "scientific" kind. Many scientists who still follow this approach today tend to consider philosophers and scientists, not to speak of theologians, of previous epochs as being naïve thinkers because they did not practice what they consider to be the true scientific research.

[95] Von Neumann 1932; 1955.

[96] Wigner 1961.

[97] Lemaître 1931a-b.

takes into account the received scientific image of the universe as being essentially stationary. However, the reaction of those cosmologists (among them we recall Bondi, Gold, and Hoyle[98]) was even more extraordinary: in order to preserve the theory of a steady-state universe, they assumed that there is a continuous creation of new matter filling the void left by the galaxies that are "moving" away from each other. It is quite likely that the authors meant such a creation as the production of random events in the sense of quantum mechanics. However, this hypothesis presented many ambiguities, left many questions open and could be interpreted as running against the fundamental philosophical and theological principle that nothing can be produced from nothing (*nihil ex nihilo*). A solid philosophical and theological background (which, as we shall see, considers continual Creation, but in a totally different sense) could have avoided a questionable interpretation.

The general lesson we would like to draw is that in critical passages of the scientific journey, scientists, for the need of generalization, are willing to put into discussion received principles and assumptions that were previously considered fundamental and to make substantial use of philosophical and theological concepts even without admitting this. The problem is that without an explicit involvement of philosophy and theology in the cognitive enterprise, these well-founded needs may give rise to many mistakes and true conceptual distortions. It is here that two new dimensions of our knowledge of nature come out in their relevance: *philosophy* and *theology of nature*. Let us first consider the former.

2.2 Philosophy of Nature

2.2.1 *Nature as a Problematic Concept*

We may wonder what precisely "philosophy of nature" means[99]. In particular, the concept of nature is a true puzzle to many, due to the enormous variety and complexity of situations that this term covers. A well-established tradition in philosophy, especially in those (post-modern) currents more involved in the so-called

---

[98] Hoyle 1948; Bondi 1948; Bondi-Gold 1948. See also Salucci 2011, §§ 3.6-7.

[99] We may assume that Aristotle with his treatise on *Physics* tried to cover a philosophy nature (a general science of becoming) as propaedeutic to particular enquiries developed in his special treatises on natural sciences (see the Introduction to the Italian translation "Fisica" edited by Ruggiu). A modern attempt can be found in Descartes 1644. One of the last theoretical enquiries in this sense is Hartmann 1950.

human or social sciences[100], has pointed out that nature is actually a cultural invention and therefore always relative to a given context (Subsec. 2.1.4). As a matter of fact, the continual progress of science displays its provisional character. For this reason, we agree with A. McGrath that "nature" cannot be considered as an immediate or unfiltered given state of affairs but involves some critical assessment. However, some scholars have been tempted to hyper-generalize this consideration, also from a philosophical point of view. In particular, the crisis of foundationalism and the weakening of the universal dimension of rationality seemed to justify a relativistic standpoint[102] (Subsec. 2.1.2) also from a philosophical point of view. Cultures as well as traditions appear to be the result of unbridgeable and mythological narrations or of different associations devoid of any reference to reality[103], the distinction between madness and sanity as a cultural artefact[104], textual interpretations as constituting the text itself[105], and finally even what we call "reality" is lost in a network of mass-media simulacra[106]. From this perspective, knowledge ultimately risks being reduced to a narration among others. This is quite surprising, because the main aim of science is precisely to go out of local and historical contexts, even if, obviously, any scientific enquiry is conditioned by specific cultural contexts[101] (Secs. 1.1-1.2). Thanks to our examination of critical realism, it should clearly appear that an epistemological assessment of science, as providing real and effective although fallible knowledge of the natural world, is still able to provide arguments against sceptical doubts and relativism. Therefore, we disagree with the aforementioned tradition that reduces philosophy to a verbal game, as it misses the crucial character of the scientific enterprise as well as of any knowledge: to try to ascertain results that have a general validity even if it is in a particular context of research that is ultimately contingent[107]. In this sense, it is fruitful to use the concept of nature provisionally understood as the world exerting constraints on our speculations (Subsec. 2.1.4) and therefore as the "arena" and the ontological context which any scientific and philosophical enquiry is concerned with. In the following we shall

---

[100] For an examination of this problem see McGrath 2001-2003, pp. 116-133.

[101] Laudan 1977; Harré 1986, pp. 8-16; McGrath 2001-2003, pp. 67-68.

[102] This risk is pointed out in the Encyclical *Caritas in veritate*, § 26. See also Polkinghorne 2009, p. 25.

[103] Lévi-Strauss 1962 and 1964; Foucault 1969; Lyotard 1992.

[104] Foucault 1972. See also White 1896, II, Chs. 15-16.

[105] Barthes 1957; Derrida 1967 and 1972.

[106] Baudrillard 1981.

[107] Peirce 1871, p. 468.

try to articulate such a notion in order to draw attention to the richness of situations and contexts of the natural world.

A relevant issue is why philosophy is necessary in this context. In other words, why should philosophy be relevant to science or even to the interaction between science and theology? The aforementioned relativistic approach in philosophy, which is unfortunately very influential, has led authoritative theologians to think that the dialogue between science and faith can be entertained without intermediary instances[108]. What we would like to show in the following is that philosophy, if led back to its most original cognitive attitude, can in fact be very helpful to such an enterprise. Moreover, the conceptual exploration of the possibility and extent to which theological statements are related with scientific work should be the object of a philosophical enquiry. Therefore, we cannot deal with the kind of problems that arise in the dialogue between science and theology without first taking into account the interrelations between the history of philosophy and the history of the formation of scientific concepts[109]. This obviously demands a drastic correction of the actual course of philosophy, and we also hope that theologians[110] and scientists will help us in such an effort. In our turn, after the epistemological assessment carried out throughout the present chapter, we shall propose (in Ch. 3) general principles of philosophy of nature with the aim of capturing the essential conceptual elements of the scientific quest for truth, while trying to avoid illegitimate philosophical absolutization of specific or provisional scientific results.

2.2.2 *Philosophy of Nature and Long-ranging Scientific Strategies*

In our perspective, philosophy of nature is essentially:

(a) Anterograde in its rational effort to help in formulating general strategies of research and preparing the ground for further advancements in learning;
(b) Retrograde in its rational effort to make the general trends of the scientific enterprise explicit on the basis of its state of development, taking into account data, observations, and experimental outcomes, including problems and difficulties in order to critically evaluate and rationally justify current scientific endeavours.

---

[108] McGrath 2001-2003, v.1, p. 41.
[109] Pannenberg 1993, pp. 32-33.
[110] Tanzella-Nitti 2010.

In both senses, we are interested in the philosophical reflection on the knowledge provided by natural sciences as a qualified expression of the human striving for nature's *integral intelligibility*. In constructing scientific theories, in fact, the purpose is not only to theoretically frame the available data and experimental results in a coherent way in order to make correct predictions, but also to reach a rational understanding of what kind of reality the available information suggests to us[111]. In other words, we are not only interested in the current practice and specific results of science, but also in considering it as a part of the cognitive journey which aims ultimately at truth (Subsec. 2.1.4). Such work deals with issues like the relationships between natural laws and the contingency of events, causality and freedom, information, matter and energy, the concepts of time and space[112].

Such an effort has very often accompanied modern science since its very beginnings and, not by chance, it has been felt by many leading scientists to be an essential aspect of their pursuit of truth. Eloquent examples of this are the various writings of such scientists, especially published during the second half of the 19th century and the 20th century, trying to spell out the nature of the scientific enterprise, its scope and limitations, its methodology, its contribution to our understanding of the world, and even to establish the underlying *"scientific world-view"*[113]. It is true that, in these works, it was physics that was the reference-discipline. However, this fact does not deprive such attempts of their generality since, for historical reasons, physics, being older than other scientific disciplines, was the first to reach solid achievements, so that it was traditionally considered as the paradigm of science and even of philosophy of nature.

A specification, however, is important here. The attempts at building a scientific world-view were *not* meant by most of those scientists as a sort of final word on what nature is. Their contribution to the dynamism of scientific enquiry may be understood rather as assessing what the main scientific results are that have thus far been reached (Point (b) above) or as having a *programmatic* value, in settling the basic scientific concepts that are helpful in fine-tuning the *strategy* that should guide science (Point (a) above). It is here that philosophy can come into play and assist the work of science. The reason is that philosophy is intrinsically *critical*:

---

[111] Pannenberg 2008, p. 9.
[112] Pannenberg 2008, p. 28.
[113] Mach 1883; Boltzmann 1905; Russell 1931; Einstein-Infeld 1938; Einstein *MW* and 1956; Planck 1965. See Tanzella-Nitti 2009, pp. 212-213.

showing that specific scientific results can never be considered to be absolutely certain and established once and for all (Subsecs. 2.1.2-2.1.4), it may hinder the crystallization of the actual status of the scientific knowledge in a closed framework. An example may help to clarify this issue. As mentioned in the introduction (Sec. 1.2), the so-called mechanist view was actually formulated when modern science was arising. This view was surely inspired by the results that the "new science" was already obtaining, but was also achieved through a critical evaluation and rejection of the "search for essences" that characterized the former cognitive strategy (since the time of Plato). As a matter of fact, such an operation was completed by *philosophers* of the Modern Ages. On the other hand, we are not claiming that philosophy should exert some sort of control on the particular activities of the scientific institutions and laboratories: this would result in a denial of the autonomy of science that we strongly support[114].

In the present age, on the basis of some of the scientific advancements and achievements of the last century, philosophy of nature is again called to perform a similar task (see also Sec. 1.3). Here, we only wish to recall three hints in this sense. Some of us have recently tried to clarify how a non-naïve realistic attitude with respect to some of the most enduring and serious problems in the interpretation of quantum mechanics[115] can make use of non-mechanistic notions like "potentiality" and "non-local correlation" that are also useful to account for recurrent aspects of physical reality at all its levels of complexity[116]. Remarkable philosophical lessons can also be drawn from biological evolution. It is, indeed, undeniable that, even independently of the assessment of its current epistemological status[117], evolutionary theory patently supports the idea that natural reality as such should be considered as fundamentally dynamical[118]. Furthermore, the huge field of neuroscience is providing more and more articulated and detailed descriptions of neural networks involved in complex organisms' fulfilment of both basic and complex functions[119]. We do think that the specificity of living and cognitive systems with respect to physical ones may consid-

---

[114] *Gaudium et Spes*, n. 36.
[115] Auletta–Torcal 2011.
[116] Auletta 2003; 2005; 2006a; 2011a, Ch. 6.
[117] See for instance Gilbert *et al.* 1996; Futuyma 1998; Gould 2002; Conway Morris 2006; Gilbert–Epel 2009.
[118] Cottier 2011; Maldamé 2011; Auletta *et al.* 2011a.
[119] Jeannerod 2006; Changeux 2009.

erably extend our understanding of nature, by leading to the formulation of more general principles and laws, thus spreading light, retrospectively, on some grounding aspects of our world.

2.2.3 *The Need for a Cross-disciplinary Approach*

Our previous examination shows that a truly *cross*-disciplinary enterprise is helpful in finding new strategies of research in such a historical conjuncture: what is needed is not some kind of juxtaposition of the fundamental results in physics, biology and cognitive neuroscience, but a true *critical* articulation of them in a comprehensive view in order to focus on the fundamental conceptual and methodological requirements. This critical assessment cannot result from specialized sciences alone, but must be carried out by making use of the conceptual rigour typical of philosophy. Therefore, the issue here is to integrate specialized scientific fields within a philosophical architecture, thus stimulating new insights that cut across the current disciplinary partition. For instance, understanding the relation between information and function in cell biology requires the contribution of many diverse fields like information and communication theory, system theory, genetics, evolutionary biology, proteomics, all of them to be philosophically integrated into a common conceptual framework. Therefore, *cross*-disciplinarity should not be confused with *inter*-disciplinarity. The latter is very often understood and practised as a limited form of collaboration among scientists of adjacent disciplines. The former aims at connecting often distant fields of study and at letting their contributions converge towards new conceptual breakthroughs; here philosophy can favour this coherent convergence, making it possible to address the manifold of the involved aspects. In other words, philosophy provides the conceptual and *strategic* framework through which particular new phenomena can be investigated, and accordingly interpreted. Summarizing, cross-disciplinarity can be considered as a trade-off between the current scientific specialization and general philosophical views.

We are fully aware that by proposing such an approach we seem to be runnig against the accepted compartmentalisations among fields and methodologies that have been the cornerstone of modern science. To avoid any possible misunderstanding on this point, let us stress that we are aware of the necessity of avoiding any confusion between fields like philosophy, science and theology, as it will also become growingly clear in the following. We are rather interested in convergence and mutual correction. We do not deny the necessity of maintaining the division among scientific fields either. The precise definition of a field of investigation remains one of the major acquisitions of Modern-Age science. Nevertheless, in some contexts it can become crucial to cross data and to exchange opinions among specialists of different fields. If this is true also for the ordinary scientific activity (especially when problems at the threshold among different disciplines arise) it becomes momentous when there is a change in long-term research strategies or when significant changes in the current strategy are in place. Let us also stress that our proposal is not completely new. Indeed, we would like to recall that this was precisely the methodology used in the *Collegio Romano*, the Jesuit Roman University that, among other things, made the reform of our calendar. We would like to mention here the name of the great Athanasius Kircher, who was called Philosopher, Mathematician, and Philologian[120]. Relying on the previous work of Gilbert[121], he opened the path to a theory of magnetism understanding that the Earth is a big magnet, magnetism is a phenomenon connected with metals and in particular with iron[122]. Moreover, he understood well that (1) there is an axis ending in the two poles of a magnetic field (which he considered as an all-penetrating or all-pervasive fluid), (2) the field propagates by starting from the two poles in a spherical form and with lines that although straight approximate the deflections that we observe today, and (3) a magnet enveloped with iron exercises its action with more strength[123]. It is clear that Kircher also considered magnetism as a *vis attractiva*[124], a statement that appears to be in contrast with the scientific methodology expressed by Newton when, referring to the occult nature of gravitation, he famously said: "hypotheses non fingo"[125]. However, that statement should be understood as a splendid example of how a great scientist is willing to accept certain conclusions (the law of gravitation) even without knowing what the cause could be and that he will not formulate hypotheses before having sufficient empirical data hinting at this cause. However, it cannot be interpreted as

---

[120] As it stands in Kircher 1643.

[121] Gilbert 1600.

[122] Kircher 1643, b. I, part I, Chs. 1-3.

[123] Kircher 1643, b. I, part II. He also proposed several experiments on magnetism. Book II is devoted to applications and Book III to the manifestation of magnetism among living creatures (see also Kircher 1667), a subject that unfortunately has become an object of scientific investigation only recently.

[124] Kircher 1643, b. I, part II.

[125] "Rationem vero harum gravitatis proprietatum ex phaenomenis nondum potui deducere, & hypotheses non fingo" (Newton 1687, *Scholium Generale*).

an interdiction to carry out such an investigation. As a matter of fact, the lack of an investigation of the gravitational force by Galilei led him to erroneously assume the conservation of circular motion since he was aware that planets have revolved around the sun for an incredibly long time[126]. Kircher's studies on the properties of light (especially diffraction) and vision as well as about acoustic waves are also remarkable[127]. Given these presuppositions, it is likely not by chance that Roger Boscovich, physicist, engineer, philosopher, theologian, and professor at the *Collegio Romano*, was able to go beyond Kircher's fluid-like understanding thus introducing the concept of field as it is currently understood, opening in this way the path to the work of Faraday and therefore to the subsequent theory of the electromagnetic field (although the idea of ether, that still survived until the beginning of the 20th century, is actually a reformulation of Kircher's conception of field as a fluid). As to this, we may also recall W. Leibniz, who was a mathematician, a physicist, a biologist, a philosopher, a theologian and a jurist. Many may take these examples to be historical curiosities and these scholars to be kinds of alchemists inclined to violate boundaries among established fields. Others, more positively, may consider them to be last examples of a sort of universal culture that has nevertheless disappeared. We think that all these kinds of judgments simply miss the reasons that have led these people to undertake the research that they did. On the other hand, what has become impossible to an individual today can be done by a team (or by a "society of minds"): as a matter of fact the scientists of the *Collegio Romano* were a school. Ultimately, to judge whether to give rise to a cross-disciplinary enterprise is possible today or not is a pragmatic matter: one needs to try and see if it is indeed possible.

Let us now examine a possible consequence of overlooking the cross-disciplinary attitude today, or at least overlooking the contribution of philosophy. Conceptions like the recent "Universal Darwinism"[128] are attempts at elaborating a conception of nature by "transubstantiating" a single discipline or even a particular theory within a single discipline into a universal explanation. For example, we stress the metaphoric abuse of the genetic theory in introducing "memes" as

---

[126] Galilei 1632, Giornata prima, pp. 25-27 and 47 (43-46 and 61-62). See the insightful commentary in Koyré 1966, pp. 257-262.

[127] Kircher 1680.

[128] Dawkins 1983; 1986.

the cognitive or cultural counterparts of genes, able to replicate themselves and to diffuse in a way that mirrors the mechanism of genetic inheritance or viral infection[129]. Since genetic inheritance is taken here as the only biological mechanism able to explain any phenotypical manifestation, this account turns out to be unsatisfactory even from a purely biological point of view[130]. Obviously, science is built through the continuous process of transposition of ideas and models from one domain to another, and even through some metaphors helping the birth of a new technical language[131] (cross-disciplinarity may currently turn out to be very useful with respect to this purpose). However, what makes the mentioned example not very scientific and even philosophically unfruitful is that the very *same* mechanisms of explanation are transferred from one domain to another, without taking into account the *specificity* of the context and the methodology that any true scientific field requires (see Polkinghorne's Point (4), in Subsec. 2.1.5). Lacking a correct evaluation of this specificity, such an approach is characterized by a mere transposition of explanations rather than providing a true *generalization* that is applicable to a vast number of particular cases, which may be seen as the quintessence of science.

Therefore, we should avoid mere philosophical restatements of the current scientific contents; philosophy cannot be reduced to a simple logical analysis of credited scientific theories. This has somehow been one of the problems of logical positivism, as it turns out to be a mere reformulation of contingent scientific results (Subsec. 2.1.2).[132] The opposite error would be to reduce philosophy of nature to pure metaphysical concepts detached from scientific research. These concepts are, of course, extremely relevant for any philosophical reflection and even represent one of the main tenets of the present book. However, to try to elaborate a philosophy of nature solely on the bases of a-priori metaphysical and ontological notions would deprive it of its "life blood", namely the consideration of scientific results. More generally, philosophy recurrently presents the risk of relying very much on speculative inferences and of assuming that the assent to pri-

---

[129] Dawkins 1976, pp. 192-195. Blackmore 1999.

[130] Auletta 2011a, Ch. 11 ; McGrath 2005, pp. 122-25.

[131] As Lewontin (2000, p. 3) says: "It is not possibile to do the work of science without using a language that is filled with metaphors. Virtually the entire body of modern science is an attempt at explaining phenomena that cannot be experienced directly by human beings".

[132] Reichenbach 1928; 1944. Grünbaum 1973. Sklar 1974.

mary facts is also a form of inference. Quite the opposite: belief in reality is primary in our experience, and this belief needs to be corrected only when it is somehow contradicted by further experience[133]. As James says, any object that remains un-contradicted is *ipso facto* believed and posited as absolute reality[134]. This grounds the idea that knowledge is error correction: in Subsecs. 2.1.2 and 2.1.4 we have also considered the temptation to reduce the relation with experience to the issue of its interpretation. So, ontology cannot be reduced to (philosophical) interpretation and, again according to Polkinghorne's Point (4), and following Bhaskar, we point out that it cannot have its subject matter apart from that investigated by science[135] (and also from theology, as we shall see), but need to be dealt with starting from what empirical sciences have to say about mundane reality[136]. Natural science is a sophistication of ordinary-experience ontology not only in postulating new entities

---

[133] See Newman's analysis (1870, Ch. 6, § 1), especially his criticism of J. Locke on this point. He correctly points out that philosophers often see the act of assent as a mere repetition of the process of inference. A similar although slightly different point is made by T. Reid (1764, Ch. 2, Secs. 3-4; Ch. 5, Sec. 3) when he rightly points out that most philosophers assume that we can have thoughts of objects independently of or before our belief in them. On this point, Hamilton (1859, I, Lect. III, p. 44) says that in the order of nature, belief always precedes knowledge as it is clear by the fact that the former is a presupposition of human education (see also Hamilton 1859, IV, Lect. XXVIII, pp. 85-87). This view is deeply roteed in the Thomistic tradition in opposition to the empiricist tradition that influenced also Kant, as clarified by Maréchal (1926, pp. 217-218). Such a standpoint, however, does not imply at all that knowl-edge is grounded on belief, neither that knowledge is an inferior ground of certainty relative to natural belief, to use the words with which Mill (1865, pp. 61-62) believed to summarize Hamilton's philosophy on this issue. Such a consequence is unjustified especially when we consider the self-correction of knowledge, a point shared by Hamilton, Newman, Peirce, James, and many others. Similar considerations hold for Mill's examination of Hamilton's statement about the unconceivability of the first data of experience (1865, pp. 75-76).

[134] James 1890, v. I, pp. 288-89; see also Auletta 2011a, Sec. 4.1.

[135] Bhaskar 1975, p. 36.

[136] This seems to have been also the tenet expressed by Aristotle in his *Metaphysics*. The tradi-tional expression "a science which studies being *qua* being" is more appropriately translated as "a science which studies that which is *qua* that which is", so that the word *qua* is adverbially referred to the verb (to study) and avoids the danger to assume that there is a special form of being apart from that studied by the particular sciences (Wedin 2009, p.126). Obviously, this leaves out the study of God's nature and of the relation between God and the world. But this is the proper object of theology, as we shall see below.

as the result of its own enquiry, but also due to its ability to make the biological, psychological and cultural background an object of its investigation[137]. Philosophy can be very helpful for clarifying these ontological problems, without claiming to solve them by logical or other formal means, or by abstract speculations. Unfortunately, the idea of an ontology treating mysterious underlying entities has spread a certain mistrust and discredit over it in the Modern Ages.

2.2.4 *Methodological Reductionism*

The Modern Ages established the *mechanistic* research strategy (Sec. 1.2). Such a strategy has proven to be extremely efficacious: actually, it has been guiding scientific research for more than four centuries. The first solid result was the birth of classical mechanics, with which the whole modern scientific way of thinking was established. This is not by chance, because science needs to start with the easiest class of problems, and only a deep understanding of such problems may eventually help to analyse more complex ones[138]. Now, mechanical systems represent exactly the kinds of problems that are treatable in the very early steps of the scientific enquiry. Mechanistic methodology was later on extended to several other scientific disciplines, proving to be very fruitful also in domains of higher complexity.

Given its successes such an established tradition risks becoming uncritically accepted. It is here that philosophy of nature can be very helpful: with its critical attitude, it may decisively concur in avoiding the uncritical reification of the current contents, which is a serious danger for science[139]. We have already considered this problem in the previous subsection. We wish to deal with it here especially in relation to two quite important philosophical issues: a) *methodological reductionism*, and b) *naturalism* and *ontological reductionism*.

Let us examine the first point. It is often heard (especially from physicists) that, among the various scientific disciplines existing today, some (e.g. physics) are "more scientific" than others (e.g. cognitive sciences and psychology). Some epistemologists even stress that physical explanations do not require other orders of reality and are therefore more basic with respect to, for instance, biological explanations

---

[137] Auletta 2011a, Chs. 12, 21-24.
[138] Carnap 1928.
[139] *Fides et Ratio*, §§ 69 and 106. See also John Paul II 1988 and Maritain 1934, pp. 96-100.

that should ultimately rely on chemistry and physics[140]. G. E. Allen may be taken as an extreme example of this attitude as he proposed to revoke Spemann's Nobel Prize, claiming that "organization" is not "a phenomenon describable in physical and chemical terms"[141]. Instead, a philosophical scrutiny of the hypotheses and theories of life and cognitive sciences shows their full scientific status, and an epistemological analysis of the experimental procedures followed in those fields clarifies that they are mostly as sound as those employed in the physical sciences[142]. The applicative reliability of life and cognitive sciences also speaks in favour of their truly scientific character. Nevertheless, we agree on the fact that the conclusions of life and cognitive sciences are less general than those of physics. This is also quite normal if we consider that living systems may be seen as a subset of physical ones, and cognitive systems as a subset of living ones. Therefore, their features cannot be in contradiction with the characters of the physical world. However, this *does not entail* their reduction to the latter. Quite the opposite, these new fields of investigation show the fundamental limitations of a reductionist methodology if taken to be the sole method possessing a scientific character. It is not by chance that a "nonreductive physicalism"[143] is proposed today, which, although conceives matter as the ultimate reality still admits the impossibility of reducing chemistry to physics, biology to chemistry, and so on.

As already pointed out, methodological reductionism, as long as it is correctly understood in its specific contributions and limits, shows its powerfulness in science in the quest for explanations as simple and universal *as possible*. Without this methodology, not only would biology be impossible, but modern science itself would never have been born. It is even a methodology widely used in philosophy although traditionally known as the method of abstraction[144]. However, the expression "as possible" above puts very specific constraints on such reductionism: scientists should try to reduce nature to the simplest explanations as far as the object of study allows them to proceed in doing this *without explaining away real problems*. For instance, we must search for an explanation of why human language has a syntax, but we are not allowed to regard such a problem as meaningless, assuming that

---

[140] See Haught 2006, p. 83; Giberson-Artigas 2007, p. 189-191.

[141] Gordon 1999, p. 14.

[142] Auletta 2011a, Chs. 15-21.

[143] Giberson-Artigas 2007, p. 189.

[144] See for instance Hamilton 1859, III, Lect. VII, pp. 122-128.

language is the immediate result of neural excitation patterns; neither are we legitimated in affirming that these patterns are to be reduced to pure chemistry. This attitude would finally result in knowing nothing about everything.

The point we would like to make is that today a reductionist methodology should be framed in a wider methodological approach in which, when problems of high complexity like those occurring in life and cognition are studied, the specific physical and chemical interactions that could be involved are also understood in the context of more complex aspects of reality, such as structures and networks, information and correlations, functions and biological goals, cognitive processes and intentional purposes. All these features should be considered as aspects of reality in their full rights, even if still linked with the material dimension: structures are always composed of physical objects, biological functions are always instantiated in bio-chemical operations, mental cognitive processes always resort to brain mechanisms. The point, however, is that structures, functions and mental capabilities have properties that cannot be entirely reduced to their material substrates either, and such properties are indispensable for understanding certain natural phenomena. Let us consider this point more carefully.

2.2.5 *Naturalism and Ontological Reductionism*

Some readers may misunderstand the previous statements and presume that we are trying to introduce a sort of anti-naturalism into science. To avoid equivocation, let us carefully define some basic terms. *Naturalism,* as we understand it, is the philosophical position that emphasizes a correct approach to the study of natural phenomena, i.e an approach covering nature in its entirety without cutting off the kinds of fields and problems mentioned in the previous subsection. However, the proponents of naturalism[145] often use such a word in different ways. Naturalism has sometimes been intended as the belief that nothing exists beyond the world available to our senses[146]; it has also been associated with materialism[147], scientism, atheism[148], mechanism and/or secularism.

If naturalism is intended in too narrow a sense, then everything would be reduced to physics and even the concept of "biological function" would become contro-

---

[145] Chomsky 1959; 1992; 1995a; 1995b; Dennett 1978; 1987; Fodor 1978. See also Auletta 2011a, Ch 20, for a discussion
[146] Haught 2006, p. 4.
[147] Haught 2006, pp. 84-85, 118.
[148] Alexander–White 2004, pp. 28-30.

versial or meaningless[149], as shown in the previous subsections. Let us deepen this problem from an antological point of view. The point lies in whether we consider biological functions as *actual natural phenomena* that are to be explained in their specificity or as epiphenomena that should be reduced to something else. Similar considerations may hold for those other concepts mentioned above – like system, information, relation, and so on – about which philosophy of nature is called to carry out an ontological enquiry. At the opposite, the fathers of modern science adopted quite a questionable ontological reductionism, which was obviously encouraged by the extraordinary successes of their reductionist methodology. Ontological reductionism may be understood as the belief that any form of reality can be reduced in principle to simple systems like the classical-mechanical ones. Nowadays, in spite of a growing body of evidence showing that things may stand in a different way, ontological reductionism still persists.

This is the reason why naturalism is often not very welcome in a theological milieu. Nevertheless, R. Numbers[150] reminds us that medieval thinkers encouraged philosophers to push naturalistic explanations of events as far as they could[151]. Now, why did Middle-Age thinkers not worry about naturalism whilst contemporary believers do? A great part of the answer (as we shall see in Ch. 4) lies in a proper understanding of God's causality and action in Creation, which was probably more studied centuries ago than it is today. The exigency that animated the medieval thinkers is still worth emphasizing: nature should be studied in its own right and science should follow its enquiry in full autonomy; even when we do not desire to put the creative and sustaining action of God into question, no supernatural explanation should be used when explaining natural phenomena in scientific terms, even when we feel that our explanations are incomplete[152]. In this sense, we support a pure philosophical-scientific understanding of naturalism as something defining the scope of scientific enquiry. This also implies that there are no a-priori limits to what should be considered as natural; and therefore the problems listed at the end of the previous subsection cannot be dropped without careful scrutiny. As we shall see, this connotation of naturalism is not incompatible with a theological understanding of nature;

---

[149] We may recall that Comte's approach (1832-40) is precisely grounded on the assumption that everything must be explained in terms of geometry and mechanical causation. For this reason he rejected metaphysics and religion as relying on final causes. See also Neurath 1931.

[150] Numbers 2003. See also Haught 2006, p. 59.

[151] See also Harrison 2007, pp. 43-46 and literature quoted there.

[152] Jou 2008, pp. 32-34.

it may even be well integrated in the latter provided that an appropriate philosophical articulation is developed.

As mentioned, we are not denying that any level of reality displays a material aspect, substrate or medium but the reasons to regard it as, in any sense, "ultimate" (according to ontological reductionism) become more and more controversial, since an enquiry strictly limited to the material substrate is insufficient for providing an adequate explanation of almost all of the phenomena of current scientific interest. As our further examination will show, it is science itself that, in its recent developments, brings us to the plausible consideration that the natural world is intrinsically characterized by "formal" constraints and interdependencies, capable of exerting causal influences on dynamical interactions. Those features have nothing magical or supernatural; quite the opposite, they constitute an essential aspect of *natural phenomena* and therefore contribute to their understanding[153]. For example, when dealing with organisms and especially with their cognitive abilities, very relevant top-down processes are implied. Top-down causation could become a (non-reductionist) major model of scientific explanation, making use of both traditional pure mechanical effects and global formal constraints: a higher level of reality can display formal constraints that are able to canalize mechanical-dynamical processes occurring at a lower level[154]. A very well known case is the endogenous production of dopamine, i.e. a process that is independent of bottom-up environmental stimuli, but is triggered by the attentional behaviour of rats exploring their spatial environment[155]. This helps us to understand properly why disciplines dealing with higher-levels of complexity, like psychology, have research programs in their own rights. So, the extended or generalized naturalism that we are proposing (which goes together with an extended concept of nature, Subsec. 2.2.1) addresses this order of considerations in the attempt at providing an integrated intelligible picture of the natural world in its actual and dynamical reality.

We will show that constraints and interdependencies are present from the basic level of reality studied by quantum mechanics up to the more complex level of human cognition. Therefore, we shall suggest that the epistemological issue of the irreducibility of disciplines should be dealt with by means of an appropriate

---

[153] See also Margenau 1984, pp. 41-42.
[154] Auletta *et al.* 2008; Auletta 2011a, Chs. 6-14 and 24.
[155] Tolman 1932; Kentros *et al.* 2004.

ontological assessment of *emergence*, basically understood as the rising in nature of more complex ways to integrate formal correlations and dynamical interactions, yet preserving a fundamental continuity across all levels of complexity.

Summing up, we think that the concept of naturalism is neutral to the possible realities that it applies to and cannot prescribe a-priori which kind of entities must exist[156]. Moreover, taking into account the previous considerations about recent developments in science, we suggest that an *updated* conception of scientific naturalism, that is consistent with all the relevant fields of current scientific research and open to its plausible future advancements, should be acknowledged with reference to the following general guidelines:

(a) The object of science is nature. Science should (intellectually and practically) seriously take into account any particular manifestations of nature, and respectfully deal with any different instances of intelligibility at all levels of complexity.
(b) Accordingly, science also has to take non-mechanistic and non-reductionist typologies of explanation into consideration.
(c) Scientific hypotheses and theories can neither be supported by, nor derived (in their specific details) from, supernatural arguments.
(d) Science cannot be considered as a sort of anti-theology or anti-religion, as this would mean falling outside of its proper scope.

We may summarize by saying that the kind of naturalism we are proposing is a *methodological* one contrary to the *metaphysical* version that implies some form of materialims or mechanism in the way explained above[157].

---

[156] Danto 1967. See also McGrath 2001-2003, v.1, pp. 126-127.
[157] Auletta-Stoeger 2010.

2.2.6 *Philosophy of Nature as an Active Partner for Science*

We have already remarked (Subsecs. 2.1.2 and 2.2.3) how useful ontological issues (and more generally, philosophical ones) can be for scientific research. Philosophy of nature does not stand in an isolated crystal-tower somehow imposing its perspectives on science, and remaining far from scientific research on the field, even if the principles that it may propose are quite general. Some (especially those who wish to reduce the scientific enterprise to its applications) might object that philosophy cannot be of any help to science because philosophical assumptions are not testable[158]. This is not exactly true. As such, philosophical statements are not *directly* testable; however, *in conjunction* with empirical scientific statements, they may not only give rise to testable consequences but also turn out to be useful in settling interpretational issues of the kind we spoke of above as well as in refining scientific concepts (Subsec. 2.2.2). It is not by chance that Heisenberg considered Bohr's work to be primarily that of a philosopher but as aiming at experimentally testing those philosophical ideas[159]. However, the clearest example of this is the famous Einstein-Podolsky-Rosen (EPR) paper on the completeness of quantum mechanics[160], a breakthrough also from an epistemological point of view, since it helps us to consider philosophical statements in a radically different way relative to the neo-positivistic framework, according to which philosophical statements, being untestable, are meaningless[161]. In the EPR paper, instead, a philosophical principle (the so called *"criterion of reality"*) is proposed: what is real is everything that we can predict with certainty without perturbing it. This principle is put at work, in conjunction with quantum-mechanical laws, to design a thought-experiment with specific empirical predictions[162]. EPR assumed that two physical systems can somehow be connected (i.e. when they are *not separated*) only through a dynamical causal interaction that at the very least presupposes the transmission of a light signal. Instead, when the two systems are separated, no local operation on one of them could result in any influence on the other. In this way, we could predict real properties of one system without taking into account the other one (without any disturbance occurring), and therefore satisfying the principle of reality.

---

[158] Carnap 1936.

[159] Quoted in Feyerabend 1969.

[160] Einstein *et al.* 1935.

[161] See, for example, Nagel 1961 and Hempel 1953.

[162] Shimony 1989; Auletta 2006c, 2007b; Auletta *et al.* 2009b.

Schrödinger[163], against the EPR interpretation, proposed that quantum systems might still be interconnected *without* any exchange of signal, thus introducing the cardinal notion of *"entanglement"* into physics, which implies the possibility of instantaneous non-local effects between different systems without any violation of relativity (i.e. without exchange of signals at superluminal speed). This is a form of interdependence without any interaction, contrary to the EPR separability assumption. The pure hypothetical framework of the EPR paper became testable when in 1960s John Bell[164] was able to derive some mathematical consequences of the EPR's assumptions (known as "Bell's inequalities") providing quantitative predictions. The experiment proposed in the EPR paper was then performed in 1980s thus ruling out the possibility that quantum mechanics is a local realistic theory[165]. Since then, many other works tried to derive testable consequences relying also on further philosophical assumptions and concepts[166]. It is widely acknowledged today that the EPR paper changed the history of quantum mechanics by starting from what many would have considered as pure speculation.

The recalled episode from the history of physics shows a meaningful way in which philosophy of nature may stand as an *active partner* with respect to scientific enterprise in contributing to the formulation of general hypotheses and educated guesses that are able to lead research heuristically along new pathways, thus shifting the dividing line between what is empirically testable and what is not. Especially whenever a deep change in natural science occurs, science shows itself to be intertwined with philosophical reasoning. In this context, it is noteworthy to recall that a great philosopher like Kant tried to establish, once and for all, a sharp demarcation between science and metaphysics, as also remarked by J. Marèchal[167]. As a matter of fact, however, some of the issues, especially concerning time and space, that Kant[168] regarded as being metaphysical in character and leading to insoluble antinomies of reason, are now object of scientific investigation[169].

---

[163] Schrödinger 1935.

[164] Bell 1964; 1966.

[165] Auletta 2000, Chs. 31 and 34-40. Auletta *et al.* 2009a, Ch. 16.

[166] Auletta-Tarozzi 2004a, 2004b. See also Tarozzi-Colagè 2009.

[167] Marèchal 1926, pp. 30-31.

[168] Kant 1787, B543-595.

[169] A. Comte was probably a follower of Kant by assuming that what current scientific methods cannot deal with should be considered as pertaining to untestable metaphysics. We recall here that Comte assumed that the chemical composition of stars is precisely what cannot be known and thus turning out to be a metaphysical statement in this sense (see also Mach's similar

2.2.7 *Fundamental Philosophical Questions*

In its continuous progress marked by finding new research fields and by dealing with new classes of manageable problems, science can and should therefore rely on the assistance of philosophy as a systematic enquiry. Philosophy is indeed able to contribute significantly to a general strategy of research, as shown already in its past history (Sec. 1.2) and in recent times exemplified by the EPR paper and its consequences just pointed out. In doing this work, it applies a critical methodology to received theories and ideas focussing on philosophical questions like:

1. The threshold of what is currently knowable and what is not,
2. The admitted forms of inference,
3. The criteria that scientific explanation needs to fulfil,
4. Foundations of science and coherence among different scientific theories,
5. The ontological status of notions like "system", "relation", "property", "function", "information", "causality", etc.,
6. The status and meaning of naturalism,
7. The philosophical issues of the monims/dualism problem,
8. The problem of the relations between contingency and necessity[170].

In the examination of the above questions, philosophy in general and philosophy of nature in particular risk being trapped in mere speculations and failing to update

---

standpoint in Subsec. 2.1.2). With regard to this, Peirce (CP 1.138) remarked: "But the ink was scarcely dry upon the printed page before the spectroscope was discovered and that which he had deemed absolutely unknowable was well on the way of getting ascertained."

[170] Point 1 has been the object of extensive debates at least in the last two centuries, as seen in the previous subsection. Point 2 corresponds to the role Aristotle assigned to logic (*An. Pr.*) and was also one of the main concerns of Peirce (1866; 1878a). Point 3 covers at least partially what Aristotle meant with scientific knowledge (*An. Post.*). Point 4 may be approximately seen as the role that Aristotle assigned to metaphysics (when first axioms are considered) on the one hand, and dialectic on the other (*Top.* 101a34-101b4), especially if we take into account that philosophy at his time also covered empirical sciences, while Point 5 seems to correspond to his idea of metaphysics when basic ontological notions are concerned (see Wedin 2009). Pont 1 is also basic to Locke's enquiry (1689), whilst Point 3 has been extensively investigated by modern scientists and philosophers (in Subsec. 2.1.2 we have already pointed out the basic literature about). Point 6 has a relative recent tradition. Point 7 has played a relevant role in the philosophical literature after Descartes' enquiry, Point 8 is typical of Middle-Age philosophy and theology and has also encountered a certain interest in the relatively recent analytic-philosophy literature.

themselves when overlooking the empirical data coming from scientific research[171]. Indeed, the range of abstract *possibilities* is always much wider than that of reality: science helps philosophy to pass from the world of mere virtuality to that of existing reality, and to consider old philosophical problems from new perspectives[172]. Indeed, these problems should be dealt with anew everytime. For this reason, they are both open and a sort of ultimate questions about knowledge and scientific methodology.

We have already considered in the previous section the third and sixth questions above. In the book we shall deal several times with the fourth question. Here, let us consider in particular the first one (with some connections with the last one). The issue raised by the physicist David Deutsch is the following: Suppose that there does not exist an all-explanatory scientific principle or at least that it is not accessible to science[173]. Such limitation would run against the aspiration of science towards universality which has been the drive behind its progress (Subsec. 2.1.6). However, if such a principle exists, it would be self-contradictory since a principle cannot explain its own origin or ground. To avoid this, we should introduce another explanatory scientific principle and so on, which leads to an endless regress (an argument found in St Thomas' development of the five proofs for the existence of God[174]). The solution of this dilemma can be found in the fact that science cannot fulfil by itself its tendency to universality (it cannot close the circle once and for ever) without betraying its own mission. However, it can find in philosophy (and in theology) the resources for dealing with this problem rationally.

2.2.8 *Search for Intelligibility*

Let us try to take one more step on the basis of the main results we have reached so far. We have stressed that science alone cannot fulfil its own striving towards the *universal intelligibility of nature*, although any scientific theory must try to cover its whole specific domain of application as exhaustively as possible by means of its own peculiar methodologies, finding its own justification as a scientific theory precisely in doing so. While each scientific discipline takes for granted its basic principles and assumptions (at least until some counter-evidence is found), philosophy, in any step of its conceptual progress, always asks about the *legitimacy* not only of the scientific enterprise but also of its own foundations. This means that

---

[171] Maritain 1934, p. 119.
[172] Auletta 2006a; 2007a.
[173] Quoted in Vedral 2010, pp. 9-10.
[174] Aquinas, *S. Th.*, I, q.2, a. 3.

philosophy is *critical* in its deepest nature and not constructive as science is (Subsec. 2.2.2; see also Subsec. 2.1.5 for the notion of critical realism), sometimes giving the false feeling that it does not progress at all[175]. On the contrary, and precisely thanks to its methodology, philosophy is able to gain and critically solidify its results even if it does so over a very long period of time[176]. Such apparent contradiction – i.e. philosophy being critical while simultaneously solidifying results – can be solved by understanding philosophy (at least in its form of philosophy of nature) as a kind of knowledge that is not anchored in a field or a specific class of problems (although it is inspired and supported by the empirical scientific results), being a sort of open journey or a practice[177]. Such a practice is positively capable of strategically helping and assisting knowledge of nature by glancing at those *big questions* that are open to a further field of investigation: theology of nature. In this sense, philosophy is a *philosophia perennis* as a dynamic effort to intelligibility not aiming at the final solution of its own fundamental problems. We appreciate the image of philosophy proposed by Hegel as progressing in a spiral-like trajectory of enquiry through which all problems are again and again submitted to critical analysis but at a growing level of complexity and specification[178]. Since philosophy is an open journey, it cannot critically account for the dimension of intelligibility in which it lives. In this sense, philosophy cannot justify itself. Indeed, to admit that there is a universal dimension of rationality and intelligibility demands an act of *intellectual faith*, that is, belief in something that can be rationally justified as an unavoidable conclusion but not fully proved or grounded. The reason is that philosophy discovers the possibility of something that is not relative but cannot be fully admitted, because, being critical, it assumes that everything is relative. In the words of H. Spencer,

> In the very assertion that all knowledge, properly so called is Relative, there is involved the assertion that there exists a Non-relative. In each step of the argument by which this doctrine is established, the same assumption is made. From the necessity of thinking in relations, it follows that the Relative is itself inconceivable, except as related to a real Non-Relative. Unless a real Non-relative or Absolute be postulated, the Relative itself becomes absolute, and so brings the argument to a contradiction[179].

---

[175] Kant 1787, Introduction.

[176] Auletta 2009b.

[177] Wittgenstein 1953.

[178] Hegel 1821, § 2. Hegel actually spoke of a circle but specified that what at the beginning of an enquiry appears to be immediate, at the end appears to be the result of a dialectic process. Therefore, traditional standpoints receive new significance after a philosophical examination.

[179] Spencer 1860-1862, p. 80.

70

With similar words, Sir William Hamilton has stressed that philosophy, in its investigation, must ultimately reach a conclusion that it considers to be impossible[180]. However, a deeper and comprehensive study will show that far from being the principle that what is inconceivable is impossible, true, it happens, on the contrary, that all that is conceivable is a mean between two contradictory extremes, both of which are inconceivable, but of which, as mutually repugnant, the one or the other *must* be true. Thus, philosophy, in demonstrating that the limits of thought are not to be assumed as the limits of possibility, while it admits the weakness of our discursive intellect, re-establishes the authority of consciousness, and vindicates the veracity of our primitive insights. Therefore, it proves to us that while we can never understand how any original datum of intelligence is possible, we have no reason from this inability to doubt that it is true. This implies in no way a sort of distrust of the power of the human reason. On the contrary, a well-grounded theological and philosophical perspective opens an indefinite path of progress to the human mind. In the words of Peirce:

> But, it is said, there must be something ultimate, something inexplicable. Undoubtedly;- in one sense. That is to say, there must be a point where each system fails; man can never attain the absolute insight. But this is not to admit that there is a point which no system can explain. Man can advance indefinitely; he cannot go to infinity, but he is able to pass beyond any assignable point[181].

---

[180] Hamilton 1859, I, Lect. II, p. 34. In his sharp criticism of Hamilton's philosophy, the great J. S. Mill (1865, Ch. 4) seems not to take sufficiently into account that also Spencer shared this point of view. On the other hand, to affirm that the notion of God is a "fasciculus of contradictions" (1865, pp. 46-48) without really examining the problem seems a bit too uncritical, especially if he affirms that we cannot say that what is inconceivabe is impossible (1865, pp. 66-68). Indeed, the main tenet of Mill is that "a conception made up of negations is a conception of Nothing", which implies that "it is not a conception at all". We acknowledge the difficulty to exhaust this problem by philosophical means alone, but we hope to show in the following that there is a theological path of investigation that may turn out to be fruitful for dealing with these issues. The important point is not to conceive this inconceivable as if it were an object in an ordinary sense of the word (1865, pp. 54-56). Obviously, we agree with Mill's criticism of the doctrine (which he attributes to Hamilton) according to which there is an immediate intuition of certain primary qualities of things (1865, pp. 13-19). It is quite possible that Hamilton (at least in his *Lectures*) understood this rather in the sense of a preliminary assumption or belief about the world characterizing the preliminary steps of knowledge, which would be much more consistent with other positions that we shall mention in the next sections. However, even if fully grounded, this criticism does not hinder us to appreciate other parts of Hamiltons' philosophy. As a matter of fact, the aim to demolish a philosophy seems to be not totally justified from the point of view of the search for truth, in that it does not contribute to any consistent and durable advance.

[181] Peirce 1866, p. 399.

A learned ignorance is thus the end of philosophy, and this is what opens to the further dimension of intelligibility provided by a theological framework[182].

2.3 Theology of Nature

2.3.1 *From Religion to Theology: An Epistemological Assessment*

Religion is a very diversified phenomenon connected to a wide variety of cultural contexts (Sec. 1.1). It seems to follow that both culture and religion have an element of arbitrariness and relativity. However, although cultures are obviously different, this does not imply that they are arbitrary (Subsec. 2.2.1). Indeed, it seems to us that any culture is equipped to solve specific problems that arise from a particular context of life. So, even if the solutions are different, they can be understood, at least to a certain extent, as optimal answers to particular challenges. We even think that it is unjustified when a "bold" philosophy raises doubts about ordinary experience in the name of the many counterintuitive results of scientific knowledge. We think, on the contrary, that there must always be substantial and specific reasons for questioning common sense as well as any received tradition[183]. Moreover, not all elements of a culture are different from one context to another, but there are also very important commonalities. For instance, studies on the early stages of human cultures have shown that there are elements born independently of each other, and nevertheless very similar[184]: the development of agriculture, pottery, the building of a State, the development of mathematics and of a computation of seasonal and astronomical recurrences. Are these examples of spontaneous convergences? We shall not deal with this problem here but ask whether or not religions also have some commonalities. As far as we can understand, many if not all religions of the world assume that the immediate (perceptual) content of our experience does not exhaust reality but there are other aspects, entities or properties that somehow constitute a sacred dimension[185] which is sharply distinguished from a profane dimension[186]. Furthermore,

---

[182] Hamilton (1859, I, Lect. II, p. 42) said that a true philosophy is the true preparative and best aid of an enlightened Christian Theology. On this point see also Tanzella-Nitti 2009, pp. 150-152.

[183] Reid 1764, Ch. 1, Sec. 5; see also Peirce CP 5.376. This is the reason why Peirce (1905) called Critical Common-Sensism the standpoint of critical realism supported in Subsec. 2.1.5, where he explicitly referred to Thomas Reid's philosophy of Common Sense.

[184] Renfrew 2011.

[185] Otto 1917; Van der Leeuw 1933.

[186] Eliade 1958.

many religions assume that there are one or more principles grounding the natural world, and that the destiny of humans as well as of other beings depends, at least in part, on those principles and related sensible manifestations of the sacred dimension. Moreover, the belief in the existence of life after death is widespread. Finally, all religions seem to assume that there is a dimension of mediation between what is natural and what is supernatural, which gives rise to rites and different religious practices, as well as also that a kind of moral discipline is important for a religious life.

Theology tries to rationalise these traditions, as well as to deal with specific relevant religious happenings that we call Revelation. In fact, we cannot deny that religion is a socially specific historical reality and can take a variety of forms and manifestations. This variety can also turn out to have irrational consequences, but whenever theological reflection is properly developed and taken into account, the presuppositions are established in order to avoid the dangers of an obscurantist attitude. This is the reason why every faith system should contain a core that has a *cognitive* value, thus ensuring a bridge to rational examination, namely to theological reflection[187] dealing with ultimate explanations of reality and therefore with truth. As J.M. Maldamé points out, theology is religion purified from mythology and idolatry[188].

This rises the problem of the relations between theology, science and philosophy. As pointed out by the authoritative theologian Paul Tillich, the first formal criterion of theology is that its object is what ultimately concerns us, and the second criterion is that our ultimate concern is that which determines our being or not-being[189], namely Creation. Although theology is not directly concerned with scientific procedures (and vice versa), it finds a point of contact with scientific research in the philosophical element that they share. According to Tillich, phi-

---

[187] Pannenberg 2008, pp. 11-12. In the words of J. H. Newman (1870, Ch. 5, § 2): "Religion has to do with the real, and the real is the particular; theology has to do with what is notional, and the notional is the general and systematic. Hence theology has to do with the Dogma of the Holy Trinity as a whole made up of many propositions; but Religion has to do with each of those separate propositions which compose it, and lives and thrives in the contemplation of them." In other words, religion is not concerned with the consistency of different statements or practices (Newman 1870, Ch. 5, § 3). It is also true that theology cannot exist without a religion as far as theology is only a doctrinal system, and therefore it displays an assent to notions, whilst religion is a body of traditions dealing with historical and experential happenings (Sec. 1.1) and therefore is utilmately based on assent to concrete objects (Newman 1870, Ch. 4).
[188] Maldamé 2003, p. 165.
[189] Tillich 1951-1963, v. 1, pp. 11-15.

losophy is indeed the cognitive approach to reality in which reality as such is the object: only the general principles, which give rise to scientific research and are reinterpreted and corrected over the centuries, are relevant to philosophy and constitute the basis of the ontological investigation that is also crucial to theology. However, one should also consider a difference between philosophy and theology: the former deals with the structure of being in itself, the second with the *meaning* and the *source* of being. The philosopher is concerned with the accordance between order and logos, the theologian with the *source* of the order and logos (and, therefore, of the intelligibility of the whole of nature). The philosopher deals with finite (secondary) causes as far as they are also objects of particular sciences; the theologian with the First Cause. However, the philosopher is also a theologian to the extent to which, by pursuing the universal logos, he cannot overlook its relation with particular rational justifications being rooted in a specific cultural background (Sec. 1.1). The theologian, on the other hand, cannot do his work without sharing the rational and critical detachment of philosophy[190].

Therefore, being theology the *rational* discipline that deals with God[191], it shows important epistemological and methodological commonalities with both science and philosophy as an enterprise that aims at finding rational explanations, formulated in specialized terminology, and developed with specific methodologies within a public research community[192]. Maldamé stresses that a good number of general methodological criteria adopted in science are shared by theology (and philosophy), namely: objectivity, accuracy in observations and terminology, careful analysis of facts, formulation of universal laws, coherence and therefore unity among different domains, search for regularities, and critical attitudes towards authority[193].

In this respect, it is also interesting to take into account the general criteria indicated by Scholz and supported by Pannenberg[194] that any proposition pertaining to a rational discipline should fulfil: (1) satisfy the postulate of non-contradiction, (2) be open to testing, in turn implying that every proposition has a hypothetical character opening the possibility to ask about its truth, (3) be part of a discipline that addresses a unified field of investigation. Obviously, the "test" here envisaged cannot correspond to an empirical trial or to a strict verification in the sense of logical positivism (e.g. tracing back theo-

---

[190] Tillich 1951-1963, v. 1, pp. 18-28 and 163.

[191] *Fides et Ratio*, §39.

[192] Stoeger 2007. See also Russell 2006, pp. 13-26 and 63-66; 2008, Ch. 10 and p. 130; Alexander 2001 pp. 242-288.

[193] Maldamé 2003, p. 155-156.

[194] Pannenberg 2008, p. 13-14.

logical statements to observations) but, as Pannenberg points out, it rather consists in the possibility of evaluating the propositions of a rational discipline according to their implications. In the case of theology, the key point is to see what the proposition stating that God is the all-determining reality, may imply (see also Subsec. 2.2.6 for similar considerations in a philosophical context). It is indeed assumed that a deep understanding of all reality is only possible in reference to the postulated divine reality[195]. Therefor, if somebody would question the rational character of theology (and *a fortiori* its relevance for empirical sciences) by pointing out that its fundamental assumptions (especially when Revelation is taken into account) are ungrounded, this would denote a deep misunderstanding about the character of inferences and even of the nature of knowledge. It is legitimate, indeed, to question the consequences of certain inferences but not their antecedents, since even questionable assumptions can lead to true knowledge, as we have shown in the two previous sections. This is already true for empirical sciences and needs to be true also for theology (as well as for philosophy). In ch. 4 we shall explore how far theology is able to give rise to fruitful and interesting consequences.

2.3.2 *Faith and Reason from an Historical Perspective*

Many still confuse theology with religion risking to forget the rational dimension[196]. An approach essentially devoted to a faithful report of the Sacred Scriptures in any detail or even to a closed *literal* interpretation of them may lead to overlooking not only the difficulties of interpreting the Holy texts[197] but also the rational dimension of theology, and may even lead to a great distortion of the religious message[198]. A European-continental trend in Christian theology (as opposed to the British tradition), especially in the reformed Churches, has considered science as being extraneous to, and even incompatible with, the theological discourse[199]. It may be worth investigating how far this theological continental tradition has in turn influenced continental developments in philosophy (an issue that we will address later on). On the other hand, it is also important to stress that the reformed Churches represented a critical stance towards knowledge that, while potentially diminishing the importance of sci-

---

[195] Pannenberg 1993, pp. 48 and 72-73; Pannenberg 2008, p. 15-16.

[196] McGrath 2001-2003, v. 1, pp. 50-60.

[197] Newman 1845, Ch. 3.

[198] Tanzella-Nitti 2009, p. 53.

[199] Pannenberg 1993, pp. 29-33 and 52-53. It is important however to stress that the original concern of both Luther and Calvin was not with scientific knowledge and its implications but rather with the human condition as being determined by sin (Harrison 2007, pp. 52-66).

entific pursuits, could also lead to a renewed attention on the conditions for augmenting knowledge and in this way favouring progresses in Modern-Age cognitive strategy[200] (see also Sec. 1.2). However, it is likely that this stance could have a beneficial influence on scientific enterprise but a less beneficial one on theological developments as such. As a matter of fact, the reformed theologians who showed a sensitivity for natural knowledge seemed to be a minority: we may recall among them P. Melanchton, J. Alsted and J. Comenius[201].

The consequence of an anti-scientific attitude is sometimes a fundamentalist approach to religion, which is nowadays the source of what is called "creationism", namely a reaction against evolutionary theories and even the notion of evolution, a phenomenon mainly spread in the United States[202]. In general terms, it seems to us that fundamentalism, unfortunately diffused worldwide and not only in a Christian milieu, pushes religious doctrines as well as believers towards irrational positions and assumptions. It is in this sense and *only* in this sense that one could reach the conclusion that faith is something blind and irrational, as for example R. Dawkins did, when he put "faith" together with "God" and the "threat-of-Hell", in the religious *meme*-pool[203]. Nevertheless, a sharp rejection of religion without any rational analysis is no less suspicious and fanatic than religious fanaticism, as Einstein pointed out[204]. Moreover, time is ripe to begin to overcome the received judgement that religion is necessarily in conflict with science. As it is well known, there is a substantial iconography stemming from A. White presenting the Galilei affair as paramounting in the warfare between Christianity and science[205]. It is not our aim to underestimate the

---

[200] Harrison 2007, p. 65.

[201] Harrison 2007, pp. 91-93.

[202] "Fundamentalism" has its roots in a series of publications titled *The Fundamentals* and published in the USA between 1905 and 1912 (see Zycinski 2006a, pp. 33-38). Some deep roots of this attitude could be traced back to that reformed phenomenon known as *Christian physics*, developed by authors like Lambert Daneau, who assumed that scriptural propositions contained scientific knowledge and possessed demonstrative value (see Harrison 2007, pp. 107-112). Very often still today people take position for or against religion by assuming that the Bible provides an alternative explanation to science, as shown by the example of E. O. Wilson reported in Haught 2000, pp. 7-8 (see also Subsec. 2.1.6). On the specific issue of the relations between evolution theory and religion, Catholicism has in general seen no particular conflict or source of difficulty as it is also acknowledged in White 1896, I, p. 82 with a certain surprise.

[203] Dawkins 1976, pp. 197-198. See also McGrath 2005, pp. 71 and 84-89.

[204] Quoted in Jammer 1999, p. 97.

[205] White 1896, I, pp. 128-157. However, many pieces of the mosaic are not recalled by White:

historical fact of Galilei's condemnation, a deprecable decision that reflects an incorrect understanding of the relations between science and theology[206]. But when such an authoritative man like J. W. Draper, a former President of the USA, writes that Science and Faith are two contending powers, the former progressive whilst the latter stationary[207] (which is also to a certain extent true since religion is based on traditions), he is probably concerned by the political or "civil" power still expressed at that time by religious institutions[208] whose worst consequence was probably the persecution of Jews and Moors in Spain[209]. To a certain extent, we understand Draper's concern. However, times have changed. Moreover, the historical evidence of a conflict between science and religion that is considered by Draper should be framed in a larger picture as we try to do in this book[210]. In other words, we ask for a reconsideration of the whole matter that can do justice to any party involved here respecting the limits of its office and prepare the path for a reconciliation that could be of benefit also for our society.

Indeed, from a historical point of view, when going back to the roots of Christianity, things may appear in a different light. Very soon in the history of Christianity the issue of the relation between religion and cultural elements coming from pagan or at least secular philosophical schools and scientific investigations

---

that Galilei received mathematical formation as well as enthusiastic support (at least at an earlier stage) by the Jesuits of the *Collegio Romano*, that he was himself a devoted Catholic, that among his objections to his detractors, one was concerned with his claimed infidelity to the Church.

[206] As it was acknowledged by the Galilei Commission, led in its final stage by Cardinal Paul Poupard: see John Paul II 1992; Poupard 1992.

[207] Draper 1875, Preface, pp. vi-vii.

[208] Draper 1875, Preface, pp. x-xi; p. 52 and especially Ch. 12. When he adds that the Church as an institution had produced harm to many people whilst science to nobody, we modestly remark that the comparison is not appropriate because science is a set of academic disciplines and also theology, being an academic discipline, has produced no harm. Institutions, instead, can always cause harm and indeed often do so, this is not only true of religious institutions. In general, the greatest atrocities come about when political powers abuse religious ideas.

[209] Draper 1875, pp. 143-148.

[210] It is clear that many kinds of bizarre ideas may be found in the history of any religion as Draper 1875 and especially White 1896 eloquently recall. Nevertheless, most of those ideas should be framed in the historical context in which they were born and often reflect the personal opinions of many, sometimes even famous authors, thus never becoming theologically credited or being subsequently abandoned. The history of science too has plenty of absurd proposals, dead-ends and unfounded constructions. Moreover, as the relatively recent history of quantum mechanics reveals, scientists can also show a considerable resistance to the introduction of new ideas.

was raised. It is true that Tertullian represented a radical position opposing philosophy by considering philosophers as the patriarchs of heretics and understanding religion as a form of radical anti-rationalism summarized in the formula: *Credo quia absurdum*[211]. However, since the beginning, Christianity, especially through the work of the Greek Fathers and St Augustine, chose to follow another path and even to consider significant parts of the pagan culture ultimately as the manifestation of God's wisdom[212]. It is worth mentioning in this context that St Augustine wished Biblical hermeneutics to take into account what, in any cultural context, are considered to be established facts. It is also true that sometimes already established facts may become obstacles when new cognitive developments arise[213], as it is evident in the Galileo affair[214]; in this case, according to St Augustine, Biblical hermeneutics should dynamically show the capability of positive innovations, as also reiterated by the Encyclical *Providentissimus Deus*[215].

Dealing with scientific knowledge is a constant also in the subsequent Christian tradition. The Christian authors who provided the bridge between antiquity and the Middle Ages (like Isidore of Seville, Alcuin and John Scot Erigena) tried to transmit factual knowledge without any use of mythology or magic[216]. It is obvious that this scientific knowledge presented many insufficiencies and shortcomings. However, we cannot charge those authors of lacking knowledge, just as we cannot criticize Lagrange for ignoring quantum mechanics or Darwin for not

---

[211] Gilson 1944, pp. 96-100. It is not by chance that Tertullian is taken by Draper (1875, pp. 39-45) to be the best expression of the essential antiscientific attitude of Christianity.

[212] Pedersen 2007, pp. 67-109. Harrison (2007, pp. 29-41) is more cautious in the appreciation of St Augustine's positive judgement about pagan wisdom. On the contrary, the judgement of Draper (1875, p. 62) is totally negative: St Augustine is said to have transformed the Holy Scripture from a guide to purity of life to an arbiter of human knowledge and therefore he was charged as being primarily responsible for the antagonism between science and religion. On these issues, see also Salucci 2011, §§ 4.4-10, and McGrath 2001-2003, v.1, pp.11-15. The latter author (pp. 36-45) also stresses the relevance of a positive interaction with the cultural and scientific environment for any Christian mainstream (Catholic, Orthodox, and Protestant).

[213] McGrath 2001-2003, v.1., pp. 60-64.

[214] Poupard 1987 and 1994.

[215] *Providentissimus Deus*, §§ 8, 15, 18, 19. See also Auletta 2007a. We may recall here that one of the first attempts at stressing the necessity of a rational and consistent interpretation of Holy texts can be found in Averroes *DD*, pp. 56-64.

[216] Pedersen 2007, pp. 129-138. See also Salucci 2011, §§ 4.11-12.

having taken genetics into account. Apart from the ripe period of scholasticism, those authors at the crossroad between Middle and Modern Ages, like Nicole Oresme, also explicitly criticized astrology and were fully aware of the theological dangers in rejecting science[217].

As announced in Ch. 1, we are now interested in finding out which theological branch is the most adequate one for dealing with natural science and philosophy of nature. We shall start by analysing certain currents in natural theology that both present serious problems and address interesting issues. Then, we shall proceed in framing our proposal in the context of the Christian theology of Creation.

2.3.3 *Natural Theology and Deism: Two Extreme Positions*

In order to proceed in our argumentation it is profitable to consider the positions that were established at the beginning of the modern scientific enterprise. In particular, we recall the momentous exchange of letters between Clarke and Leibniz on several scientific, philosophical and theological issues[218], where two clear-cut positions emerged:

-   That of Clarke (and of Newton) according to which nature is not and cannot be provided with the means to be self-organized, showing even a natural tendency to disorder and degradation (which is a genial anticipation of the second law of thermodynamics), so that the Creator recurrently assists the universe in order to establish the natural connections between bodies in terms of forces and organization principles. This is the ground of what is traditionally called *natural theology*.
-   That of Leibniz is paradigmatically expressed when he wrote that Creation was empowered with all elements and features necessary to its autonomous course without any particular assistance by God, apart from the initial *fiat*. This is the ground of Deism. Indeed, Leibniz supported the idea that the world can be exhaustively explained through mechanical causes even if there is a full correspondence (a pre-established harmony) between mechanical and final explanations[219]. Moreover, he seemed to interpret the latter as representing the

---

[217] Pedersen 2007, pp. 212-214.

[218] Leibniz-Clarke 1715-1716; Leibniz 1689a.

[219] It is true that he sometimes says that mechanical explanations are insufficient and should somehow be integrated with finalistic ones (Leibniz 1710-1712, § 11). However, he very often insists on the full autonomy and self-sufficiency of the former (Leibniz 1712-1714, §§ 79-81)

kingdom of Grace[220] and therefore as not being natural causes at all. However, there is also a crucial difference between Leibniz and most Deists. While the latter proposed a vision of the cosmos as something absolute (and here they rather share Newton's conception of an absolute space and time), Leibniz's standpoint was relational as the universe displays a huge web of relations such that everything is correlated with everything else[221].

There are also significant common tenets between these two standpoints. Both have tried to ground theology upon the scientific results of the mechanist science of that time (Subsec. 2.2.3). They had only *interpreted* these results in different ways: natural theology saw in the mechanical organization of nature the product of contrivances and design requiring a special assistance of a demiurgic God moulding an unorganized matter; Deism (finally connected with the rise of rational mechanics in France) saw in the mechanisms of nature the expression of the eternal and necessary laws ruling being.

Natural theology has two main variants: the first one deals with God's providence on the overall evolution of the universe, and this represents Newton's original position. The second variant maintains God's action in the created world especially as directly and specifically determining *certain aspects* of the on-going of nature, in particular when science seems to fail in understanding the actual complexity of a widespread number of natural phenomena (e.g. living beings). Here we shall focus on the latter approach, and shall come to the former later on.

Natural theology, although having some loose connection with Middle-Age theology, was born in the English intellectual world during the 17th and 18th century[222]. W. Paley's work may be considered as the paradigm of the British natural theology. Paley's *Natural Theology*[223] begins with a comparison between a stone and

---

and criticized Descartes precisely because he introduced a form of influence of the mind on the body (Leibniz 1686a). It should also be considered that he was a supporter of biological preformationism (Leibniz 1710-1712, § 6; 1712-1714, §§ 74-77).

[220]Leibniz 1710-1712, § 15; 1712-1714, § 87.

[221] Leibniz 1710-1712, § 3; 1712-1714, §§ 56 and 61. See also Subsec. 2.2.6.

[222] McGrath 2001-2003, v.1, pp. 241-248; Tanzella-Nitti 2009, p. 77. See also Newman's opinion quoted in Hughes 2009, p. 209.

[223] Paley 1802. For a careful analysis see Ayala 2007, pp. 15-26. In this book, Ayala (p. 2) makes a strict parallelism between Paley's approach and St. Thomas' fifth way. We, however, shall argue in the following that there is also a difference between them.

a watch found during a walk. According to Paley, the presence of the stone on the ground is fully justified through natural explanations: it is a pure aggregate of matter that has been formed and displaced there by means of rough natural forces. On the contrary, an examination of the watch will lead to the conclusion that it is an artefact built for certain specific purposes by somebody having a determined design. The watch, for the general understanding of that time, was a good metaphor for dealing with living beings (animals and plants), whose examination constitutes the largest part of Paley's treatise. The author is particularly interested in both mechanical (like animals' muscles) and non-mechanical (like chemical) processes, drawing the conclusion that in both cases the action of a skilful contriver is necessary.

Nowadays, intelligent design[224] may be seen as a follow-up of natural theology according to the second variant mentioned above, upholding that certain characters of nature may, and even should, be explained by admitting the intervention of a (often not better specified) supernatural intelligent cause responsible for their appearance at specific points of the world history[225]. It is worth mentioning that the major representatives of intelligent design are scientists committed to religion but not professional theologians, implying a certain immaturity of their mode of argumentation in these issues. R. Numbers reports that the birth of the recent organized movement of intelligent design originally stemmed from a reaction to the anti-religious approach of many Darwinists in the United States. However, he also notes that such movement soon enlarged its positions to an epistemological thesis claiming the *scientific* legitimacy of supernatural explanations whenever evidence warrants them[226] (Subsec. 2.2.5).

A comprehensible exigency lying beneath such a trend is to obviate a complete detachment of God (or whatever supernatural intelligence) from the created world. Apart from the claim of the scientific reliability of their statements, an enormous problem is that intelligent-design supporters end up in implying the idea of the "God of the gaps", that is, of a God called up to explain specific aspects of the world that a proper scientific explanation (at present) cannot account for[227].

---

[224] Behe 1996. Dembski 1998; 1999. Johnson 1991.
[225] For an insightful account of this matter, see Ayala 2007, pp. 137-160.
[225] Numbers 2011.
[227] Haught 2006, p. 67. See also Russell 2008, pp. 126-127.

An extreme consequence of the latter point is that intelligent design would deny chance or contingent aspects to play an important role in the created world, which is actually in sharp opposition to what science is increasingly discovering about nature. Interestingly, Pannenberg notes that the introduction of chance into our understanding of nature, typically represented by the influential Monod's *Chance and Necessity*, was originally aimed at destroying any argument from design, which actually came some time before the explosion of the intelligent design movement. Nevertheless, ruling out the arguments from design by means of the argument from contingency does not imply at all a denial of the continual creative, sustaining, and governing activity of God. Rather, as Pannenberg also stresses, in the contemporary theological interpretation of nature, the element of chance or contingency is even more important than design, since contingency and the *emergence of novelty* harmonize very well with the biblical view of God continuously supporting nature and humankind[228].

Deism, on the other hand, may be seen as maintaining that God refrains from intervening in the world He created, letting it proceed according to the order He conferred at the "*moment*" of Creation. It is not by chance that Deism considered Revelation to be an immediate, natural, and universal happening whose modality is Creation itself[229]. In other words, Deists attributed no relevance to any form of historical religion (Subsec. 2.3.1) but assumed that the true religion was given to any rational being through rational and natural means from the beginning of the world. Thus, they treated religion as a *theory* (Sec. 1.1), so that theology, philosophy and science finally coalesced into a single rational system of knowledge[230]. In this way, the natural world and rationality became wholly self-sustained, and no particular intervention of God was needed towards any created being, humans included. At most, Deists acknowledged that humankind (as well as individuals) needs a period of education in order to attain a full rational understanding of nat-

---

[228] Pannenberg 1993, pp. 46 and 57; 2008, p. 46. See also Haught 2000, pp. 6-7.

[229] Ray 1727 and Tindal 1730.

[230] We may recall here J. H. Newman's criticism of those who believed that religion could be secured and sustained in the mass of men by acquaintance with literature and physical science, and through the instrumentality of Mechanics' Institutes and Reading Rooms (Newman 1870, Ch. 4, § 3). He also warned not to attempt in philosophy what was once done so in religion. Indeed, the ascendency of faith may be impracticable, but the reign of knowledge is incomprehensible.

ural religion. This implies that God is often understood as a Pedagogue, and in this way a deep meaning of Creation as well as the issue of Redemption and Resurrection are completely set aside, depriving the world and humanity of any soteriological and eschatological dimension. In this context, we recall that Deism is characterized by an understanding of the world as a fully intelligible mechanical engine. This is not compatible with any religious belief, which indeed supposes a sense of the deep mystery of our universe and of the unavoidable inadequacy of our knowledge. The latter attitude deeply motivated the philosophical thinking of Plato and Aristotle and is still vividly present in modern scientists like Einstein[231].

We think that the difficulties affecting both Deism and intelligent design may be solved by means of a deepened and more sophisticated way of understanding the relations between God and the created world, with particular reference to the distinction between the First Cause and secondary causes as well as to the theme of *Creatio continua*; all of these issues were already elaborated within the context of medieval theology of Creation, as we shall see in Ch. 4. For the time being, we would like to stress that the traditional Christian approach to these matters is known as the analogical way to understand the relation between God and Creation. We shall try to understand the world as the text in which *vestigia trinitatis* (i.e. the signs of the Trinity in the created world) have to be found and humankind as *imago Dei* (i.e., the image of God). In this sense, we try to avoid both the alternatives of (a) a God interfering with the natural course of Creation, and (b) a complete detachment from one another.

In this context, we wish also to emphasize that the young Darwin himself was deeply impressed by Paley's work, and his theological reflections on the aspects raised there represent an important background for his own intellectual development leading to the mature theory of evolution. We see again that theology can represent an important background for scientific developments especially when scholars are not satisfied with easy solutions. Obviously, another issue is whether or not theology is subsequently able to integrate in its framework results that have been found even overcoming some of its previous assumptions. As we shall argue in the following, theology is perfectly capable of (1) preserving the freedom of the scientific enquiry in any natural domain without supporting a separation between God and the created world, (2) including at least some of the scientific re-

---

[231] See Jammer 1999, pp. 73 and 81-82.

sults in its horizon, (3) affirming the *reasons* underlying personal faith in God as long as the human cognitive enterprise is considered in its actual integrity, and (4) maintaining the idea of a providential plan of God especially addressed to the salvation of humans and to the intelligibility and beauty of the created world, without implying any destiny of nature or a demiurgic moulding of matter, as it was clearly pointed out by St Augustine.

2.3.4 *Back to Mythology?*

Whenever established scientific paradigms are no longer capable of giving appropriate answers to the questions that science itself raises, then not only new paradigmatic candidates arise to guide future research (as T. Kuhn[232] pointed out), but also the temptation of abandoning scientific *mentality* as such and, with it, the search for an objective and testable knowledge, gets stronger. The result is the flourishing of a number of cultural movements featuring an often uncritical syncretism of issues ranging from vitalism to astrology, from numerology to magic beliefs, from the extemporaneous retrieving of ancient myths to the commixture of the latter with modern or post-modern cultural fashions, etc[233].

Some of the current cultural developments in our societies show indeed significant *mythological* features. A mythological conception usually concerns the claim, the belief or even the hope that the divine power intervenes *physically* in the world, as a cause among (and therefore contrasting) other causes, in order to solve our problems (whether cognitive, existential, or related to our own welfare) as well as in the specular ability of humans to elevate themselves to divinity becoming half-gods partaking in that power[234]. This implies that mythology is always concerned with the sacralisation of a specific aspect or reality of the world *at the expense* of other ones (*This* emperor or dynasty or even family having a divine origin or status), running in contradiction with a fundamental character of Creation: diversification and plurality which witnesses God's love as extended to everything. We live indeed in a rich and fertile[235] universe, in which several so-

---

[232] Kuhn 1962, Ch 7.

[233] See the document of the Pontifical Council for Culture on "New Age", at: http://www.vatican.va/roman_curia/pontifical_councils/interelg/documents/rc_pc_interelg_doc_20030203_new-age_en.html

[234] See Salucci 2011, Ch. 1.

[235] Coyne 2005.

lutions pop up from a common ontological root and often converge into a common horizon of intelligibility. Without overlooking the soteriological and eschatological character of the original message of Christianity, we may retain that a relevant part of the problem rests in the way we understand the modality and the consequences of divine action. Whenever the divine is expected to intervene in the physical world, satisfying our desires, there is the risk of confusing spirituality and elevation (which demands humility) with the culture expressed by epic tales or cosmogonies.

The problem is so difficult that also brilliant authors who deeply devoted their lives to both science and religion could not settle it adequately. Teilhard de Chardin was not only a distinguished palaeontologist, but also one of the few scientists and priests who tried to propose the complementarity of Darwinian biological evolution with Catholic tradition at a time in which most believers found difficult to fully assimilate the theory of evolution. In this way, he helped the Catholic Church to accept the idea of evolution. Indeed, we are indebted to such a bold attempt, in which he provided remarkable philosophical suggestions relative to evolutionary biology as, for instance, the idea that during evolution it is possible to observe a progressive detachment of organisms from a totally passive relationship with the environment, so that the mind and a spiritual dimension can emerge[236]. In this way, Teilhard stressed the immense potential value of the theory of evolution not only in itself but also for other fields of investigation: the openness of creation to the future[237]. Yet, not being himself a professional theologian (although a talented religious thinker and powerful mystician[238]), he ended up in attributing both a cosmic and a soteriological dimension to biological evolution and composing a narration in which theological and scientific instances turn out to be mixed up[239]. In this way, both theological and scientific methodological requirements as well as competences are not respected; it is therefore possible to uphold the wrong conclusion that biological evolution is steered by a divine intervention and that therefore religious issues could determine a scientific research program. As a matter of fact, Teilhard de Chardin considered evolution as converging to a final destination (the

---

[236] Teilhard de Chardin 1955. See Tanzella-Nitti 2009, pp. 7-8, and 2001.

[237] As stressed in Haught 2010, p. 138.

[238] Teilhard de Chardin 1923.

[239] See also Salucci 2011, § 4.16.

so-called "Omega-point") that he considered to be deeply interconnected with Christology. As recalled by Pannenberg[240], Tipler tried, in a similar way, to introduce the ultimate convergence-point of cosmic evolution directly into the context of physics, which could represent a source of confusion between God's first Creation and the Kingdom of God, like in myths of progress towards a final utopic state of humanity[241]. It is also true that Teilhard's and Tipler's positions may be regarded as highly sophisticated when compared to the current (almost countless) attempts at giving rise to what we may call, in general, "modern-times cosmogonies".

On the contrary, science and theology should converge by applying a de-mythologization strategy[242]. We recall that the essence of the Biblical message is anti-mythological[243], and that medieval Christian theology used to be a de-mythologizing intellectual activity expressed in the proposal of a philosophical naturalism, as we have seen above (Subsec. 2.2.5). Indeed, a correct philosophical naturalism blocks any attempt at mixing up religion and science. This obviously did not (and does not) mean that the critical philosophical activity is to be seen as the all-encompassing way of thinking, ultimately allowing to get rid of the mysteries and the dogmas coming from divine Revelation. On the contrary, truths of faith may be better acknowledged and even testified, whenever our best intellectual efforts are directed to them, thus letting us face the natural limits of our comprehension (Subsec. 2.2.8) and, simultaneously, recognize the reasons to accept them as a gift from God's Grace[244].

---

[240] Pannenberg 2008, p. 27.

[241] Wright 2007, pp. 94-100.

[242] Pedersen 2007, p. 119.

[243] Bultmann 1941. John Paul II (1988) recalled that Mesopotamic mythologies were purified in order to be incorporated in the book of Genesis. We can then accept the word "myth" when it denotes the allegoric or metaphoric forms used by many cultures, but not when it is used to refer to the essence of the Scriptures. When a literal reading separates the language that is rooted in a specific culture from the true meaning, we have mythology in the bad sense of the word (see Tanzella-Nitti 2009, p. 29).

[244] Benedict XVI, *General Audience 23 June 2010: Human intelligence facing the mysteries of Faith* (3rd Catechesis on St Thomas Aquinas).

2.3.5 *An Example of Fruitful Linkage between Theology and Science*

Let us now consider a well-known relatively recent episode in the history of science. Famously, Einstein was convinced that God does not play dice with the universe[245]. This statement was expressed as a claim against the "genuine chance" that the widely accepted interpretation of quantum mechanics assigned (and still assigns) to microphysical systems. The question we would like to raise is the following: did Einstein's convictions on theological matters influence his *scientific* work? The EPR paper (Subsec. 2.2.6), which was explicitly written to disprove the aforementioned interpretation of quantum mechanics, did not address theological issues at all, and the arguments developed therein were never understood to be theological. As mentioned, the paper had (and still has) deep consequences in the furthering of physics, so that its scientific import cannot be put into question.

However, Einstein's distaste for the introduction of genuine chance within a general world-view (that also represents the motivation of the EPR paper) very likely finds its roots in his religious background, within which genuine chance is unacceptable[246]. Einstein was accustomed with the Jewish tradition and, in his faith in determinism, shared a common ground with Spinoza's understanding of Nature[247], even if we cannot say that he had been directly influenced by the latter[248]. In Einstein's world-view, God appears as Nature's legislator[249]. It is also true that Einstein did not follow Jewish religious rites and traditions and that he rather embraced a sort of cosmic religion in which there was no place for a personal God. However, this standpoint also has a certain Jewish background, in particular in the interpretation of the second commandment, as it can be found in the work of Maimonides[250]. Moreover, in his later years, Einstein seemed less categorical in devaluating religious traditions[251]. In any case, he has strongly supported the idea that religion and science

---

[245] Einstein 1926.

[246] See Jammer 1999, p. 58.

[247] Spinoza 1677.

[248] Jammer 1999, pp. 44-45 and 144-149.

[249] Jammer 1999, pp 122-123.

[250] Maimonides *GE*, I, § 56; III, §§ 29-30. See also Jammer 1999, p. 74. Sometimes, Einstein seems to deal with (and to reject) only a mere literal interpretation of the Bible. This could also have a certain root in Maimonides' work, as shown in Jammer 1999, pp. 141-142. Unfortunatley, the issue of a personal God and that of the literal interpretation of the Bible are very often mixed up.

[251] Einstein 1937; 1939. See also Jammer 1999, pp. 114-123.

are complementary[252] and even that the former is the main source of the latter[253]. In this sense, his religious ideas do not merely have a personal relevance, as it is the case e.g. for his political convictions[254].

This religious background did not determine the specific hypotheses and arguments proposed by Einstein in the EPR paper[255], neither was it ever introduced as providing arguments for his scientific conclusions. In fact, it inclined him to ask a question in a radical and fundamental way: is our universe truly random or not? These kinds of questions are indeed not connected with a specific domain of investigation, even if, obviously they were ultimately addressed at solving a problem e.g. arising from quantum mechanics. Indeed, the issues of Einstein could be reformulated in fundamental scientific questions: is nature ruled by laws that have an absolute value? Do these laws allow specific predictions about any future evolution of our universe? Has reality got an objective character that is independent of the human mind? And, foremost, is the order of nature something absolutely objective? When he came to express these issues in scientific terms, his religious tenets were no longer explicitly formulated. However, the influence of the latter on the intellectual deepness of those questions should not be denied. Similarly, it is well known that N. Bohr, the main adversary of Einstein in this discussion, who most coherently supported the idea that there is no objective reality independently of the experimental context (and even perhaps of the human mind) and argued that there can be breaks in the causal chains (or at least certain situations to which causality does not apply), was heavily influenced by the theological-philosophical positions of S. Kierkegaard, I. Kant, and W. James[256].

In other words, we are not claiming that theology as such should determine or even produce scientific programs or ideas, neither that theological arguments should be, in one way or the other, relevant for the solution of specific scientific problems. Rather, a theological frame can help us very much to formulate more radical questions and to put scientific ideas on a clearer background[257], helping us to avoid pos-

---

[252] He said (1939, p. 24): "science without religion is lame, religion without science is blind". See also Jammer 1999, pp. 93-95; Jou 2008, Chs. 1-2.

[253] Einstein said (1939, p. 24): "Science can only be created by those who are thoroughly imbued with the aspiration towards truth and understanding. This source of feeling, however, springs from the sphere of religion". See also Jammer, p. 55.

[254] See Jammer 1999, p. 38.

[255] Eistein *et al.* 1935.

[256] Bohr 1928; Murdoch 1987, pp. 225-235.

[257] Interesting reflections on this issue can also be found in Pannenberg 1993, p. 80. See also Russell *et al.* 2008.

sible misunderstandings or confusions about the scope and significance of the kind of explanations that scholars provide (Sec. 1.2). It is well known that Einstein was ultimately wrong about his interpretation of quantum mechanics[258]. Nevertheless, the EPR paper influenced a significant part of the history of physics thus far, so that it can be said that most of the developments in quantum mechanics in the latter forty years are a consequence of the questions treated therein. Thus, it may be interesting to raise the question why an erroneous interpretation could be so fertile (Subsec. 2.1.5). When Einstein raised the big question about the randomness of our universe and finally rejected the possibility of chance, he started from the assumption of a radical intelligibility of our world in which no random events can happen. We may say today that his vision about the intelligibility of our universe was too narrow (and in the following we shall provide reasons for supporting a different view). However, the fact remains that this striving for intelligibility[259] and Einstein's sense of the beauty and simplicity of our universe as instantiating an order that does not depend on us, came out of his deep religious conviction[260], and is precisely what makes the difference and confers to such a work its intellectual and cognitive power able to set in motion a whole development of physics. Therefore, theology, and especially theology of nature, takes or can take part in the cognitive search for a radical intelligibility of our universe[261], ultimately striving to the sacred goal of truth[262] (Subsecs. 2.1.4 and 2.2.8). By paraphrasing the words he used once when speaking of Newton[263], we can say that Einstein was "a scene on which the struggle for eternal truth took place".

---

[258] Auletta *et al.* 2009a, Ch. 16.

[259] Einstein said that after the first period of his life in which he was under the influence of the empiricist ideas of Hume and Mach, the problem of gravitation converted him "into a believing rationalist, that is, into someone who searches for the only reliable source of truth in mathematical simplicity" (Letter to C. Lanczos, 24 January 1938, quoted in Jammer 1999, p. 40).

[260] See also Hoffmann 1979.

[261] Lonergan 1957, pp. 696 and 706; Stoeger 2007, p. 230.

[262] Einstein 1937. This is the reason why Einstein thought that the goal of knowledge, which integrates theological and scientific aspects, and therefore was incarnated in the admirable unity of ecclesiastical and secular cultural institutions, is sacred (1937, p. 7). For this reason, he deprecated the climate of growing hostility between ecclesiastical and secular schools of the 19th century. With similar words, Peirce (1898a, p. 29) says that in philosophy (where we are not occupied with utilitarian tasks) science is sacred to us.

[263] Eisntein 1942.

2.3.6 *Theology of Creation and Theology of Nature: General Guidelines*

As far as we are envisaging a new integrated cognitive enterprise that aims at articulating scientific explanations with philosophical investigation and theological interpretations about the universe, it would be almost obvious to address ourselves to what, particularly in a Catholic environment, was traditionally called "theology of Creation"[264]. Two meanings of the word "Creation" should be taken into account[265], that is, the continuous act of God in creating the universe and the universe itself as created by God[266]. Our exploration of the interactions between theology and natural sciences is confined to this second perspective,[267] taking advantage of the mediation offered by what we have proposed as a cross-disciplinary philosophy of nature[268] (Subsec. 2.2.3). Moreover, when dealing with Creation we should distinguish between the created world in which we currently live (the first Creation, whose core, theologically speaking, is the Incarnated Christ) and the Kingdom of God (the new Creation) announced by Jesus and his Apostoles[269]. It is clear that a Theology of Creation needs to take into account these two aspects (and therefore needs to rely on Revelation). In this book we shall deal with the first Creation only, in a particular respect without claiming to provide any original contribution. We could call our enquiry "rational theology of created nature", *theology of nature* in short, following an original suggestion by Pannenberg[270]. The latter is surely to be considered as a part of the traditional theology of Creation but with a specification: the expression "theology of nature" stresses that one of its starting point is represented by those elements that are the object of current scientific theories, although re-elaborated, interpreted and generalized by the kind of philosophy of nature sketched in Sec. 2.2. Avoiding any vi-

---

[264] Salucci 2011.

[265] Tillich 1951-1963 v. 1, pp. 204 and 252-254.

[266] Tanzella-Nitti 2009, pp. 18-19.

[267] See also Guessoum 2011, p. 35 for a similar perspective in the Islamic world. Interestingly, some Islamic authors have stressed that there are three sources of knowledge: senses, reasoning, and Revelation (Guessoum 2011, p. 55).

[268] See also Pannenberg 2008, pp. 28 and 91 for clear statements in favour of these general epistemological guidelines.

[269] Wright 2007, pp. 78-79.

[270] Pannenberg 1993, p. 72. See also Tanzella-Nitti 2009, p. 21. By distinguishing between theology of Creation and theology of nature, we hope to take seriously into account the arguments developed in McGrath 2001-2003, v.1, pp. 134-138.

talistic implication, a philosophical perspective seems necessary in order to provide an alternative to both deism and intelligent design (Subsec. 2.3.3). The other source of the theology of nature is represented by the conclusions of the more general theology of Creation, which, as mentioned, finds its inspiration in the Revelation. In sum, theology of nature may be understood as a chapter of theology of Creation. Accordingly, our attention is primarily focussed on the universe as created by God that can be designated by the word "nature" (Subsec. 2.2.5). "Nature", in this sense, should be understood as a formal specificity that characterizes the context in which finite beings arise and interact[271] (Subsec. 2.2.1).

It is well known that Karl Barth was concerned with a possible chiasm between a rational theology and a revealed one. However, this concern should not exceed an understandable appeal of not overlooking the Biblical sources of God's self-Revelation and deny the legitimacy of a rational theology as such[272]. On the other hand, it is also true that theology cannot detach the descending and revealed image of God from the human ascending search for truth[273].

2.3.7 *The Big Questions*

Theology of nature is the proper place where the "big questions" about the natural world arise[274]. Such a kind of questions should be regarded as the natural and indispensable completion of scientific and philosophical rationality in its intrinsic effort toward universal intelligibility. To take just one example, understanding that the scientific question about the Big Bang and the original quantum-field fluctuation are of a different kind with regards to the philosophical and theological question about " Being and Nothing" helps scientists, philosophers, and theologians to clarify the nature of their respective fields of enquiry: indeed a vacuum is not nothing, for it is a possible state of a field and therefore presupposes the existence of that field[275].

---

[271] Tanzella-Nitti 1997 and 2004

[272] McGrath 2001-2003, v. 1, pp. 264-286.

[273] See also Tanzella-Nitti 2001.

[274] Among others, J.F. Haught (2006, pp. 19-20) clearly states that it is the business of theology to provide ultimate explanations.

[275] Barrow and Tipler 1986, pp. 440-444. See also Polkinghorne-Beale 2009; Tanzella-Nitti 1994; 2002.

Some of the main questions that seem to be relevant to us for the kind of theology that we are envisaging are[276]:

1. Why there is something rather than nothing[277],
2. The theological aspects of contingency and its relationship with necessity[278],
3. The problem of infinity[279],
4. The features of the finite and temporal being and its relations with the infinite and a-temporal being,
5. The relations between the principles of philosophy of nature and God as the ultimate principle,
6. The theological aspect of the monism-dualism issue in nature.
7. God's mind as the ultimate ground of the intelligibility of nature, the question already raised in philosophical terms in Subsec. 2.2.8.

We emphasize here the fundamental difference between these kinds of questions and the philosophical issues raised in Subsec. 2.2.7. In spite of the general character of philosophical investigations, they are addressed to specific classes of (epistemological, methodological, logical, and ontological) problems and do not deal with fundamental questions like the origin of being or the meaning of the universe that are typical and unavoidable for theology.

It is also worth noting that the first question (Why is there something rather than nothing?) certainly pertains to philosophy, in particular to metaphysics. Only that the solution to this problem cannot be found through philosophical means alone but requires a theological approach, that is, an approach that understands being as received, manifested or revealed. This goes much further than the critical methodology of philosophy. In Heidegger's language (which to a certain extent aims to substitute a sort of post-metaphysics to traditional theology), the chief part of philosophy that we call metaphysics deals with all forms of specific beings, whilst a more radical approach attempting at overcoming metaphysics would deal with the Being. However, this could only be done if Being reveals itself, if the Being's

---

[276] Following somehow Pannenberg 1993, pp. 15-27. All these points have attracted attention by both Middle-Age and Modern-Age theologians.

[277] Leibniz 1710-1712 §; *PS*, v. 7, pp. 289-291.

[278] Torrance 1985a and b.

[279] Barrow 2005.

relation to the essence of man happens[280]. The problem is, obviously, to which extent this replacement is legitimate or fruitful, an issue that we shall touch later on.

It is interesting to remark that S. Hawking raises similar ultimate questions, in particular[281]: Why is there something rather than nothing? Why do we exist? Why this particular set of (natural) laws and not other ones? However, he thinks that philosophy (and even more so theology) is dead and has been substituted by science, which therefore candidates itself to a sort of ultimate explanation of everything. While we think that the fact that a famous scientist dealing with these kinds of questions shows how relevant they are to science and that science cannot be confined to the examination of specific problems only, his proposal to candidate science to take the role of philosophy and theology appears to be inappropriate (Subsec. 1.1.6), and we shall also consider some consequences of this point of view.

2.4 Summing Up

In conclusion, let us schematically summarize the relations among the natural sciences, philosophy of nature, and theology of nature according to what we have said so far.

- Science is *potentially* a total, all-embracing rational enquiry, but is always specialised *de facto* and therefore confined to well defined scopes. The natural sciences, in all their multifaceted specializations, provide the basic contents on which a philosophy of nature should be built in a truly cross-disciplinary fashion, thus giving rise to a *naturalistic* world-view not committed to ontological reductionism or materialism. In its fallible search for truth, science is intrinsically open to philosophy. Moreover, such a naturalistic world-view induces theology of nature to keep itself updated to the most recent scientific achievements.
- Philosophy of nature may offer new perspectives to the natural sciences and interpretative hypotheses that are at least potentially testable, thus promoting new research strategies. Philosophical questions are differently addressed in time according to the dynamical relation that they have with science. Philosophy carries

---

[280] Heidegger 1928-29; 1955. Heidegger considers theology as not being sufficiently radical, as too much dominated by apologetic concern. This is an issue that must be seriously considered by theologians.

[281] Hawking-Mlodinow 2010, pp. 5-10 and 29.

out this work with an openness towards the fundamental questions raised by theology. This dynamism allows philosophy to avoid a sterile and relativistic universal criticism and consequently to establish a grounded bridge between empirical research and theological investigation.

- Theology of nature provides an interesting rational and overall context for science and philosophy by showing intelligibility as the ultimate ground of the consonance between scientific theories and the world. It performs this work by taking into account both the autonomy of other disciplines and the human universal striving for knowledge[282].

---

[282] *Aeterni Patris,* §§ 29 and 39.

3

PRINCIPLES OF PHILOSOPHY OF NATURE: A PROPOSAL

Due to its intrinsic character, philosophy of nature has to start from very general assumptions that are a sort of conceptual refinement of some essential achievements coming from scientific and empirical enquiry. We would like to clarify that, although there is a certain under-determination of philosophy (and theology) of nature relative to the natural sciences, the current results of the sciences together with their established principles and laws set clear constraints on the possible character of philosophy (and theology) of nature[1]. For instance, a philosophy of nature centred on a kind of fixism of biological species, mechanism, determinism, or generalised teleology as proposed by the supporters of natural theology is no longer tenable (Subsec. 2.2.3). This is due to the fact that science, by means of its thorough investigation, has shown that the natural world, at nearly all its levels, simply does not correspond to those simplifying generalizations; therefore, if we are interested in proposing and maintaining a philosophical framework *consistent with the scientific knowledge of nature*, we are forced to abandon them. In this way, science represents a sort of body of negative evidence for philosophy, but cannot ground positively the latter[2] (Subsecs. 2.2.2-2.2.3).

Consequently, in the following we shall try to propose three putative and heuristic principles[3] for an updated philosophy of nature that are strongly rooted in the current scientific knowledge of nature. As we shall see, they do not correspond to any principle or law characterizing present-day scientific theories; rather, they are cross-disciplinary principles in the sense that they have been drawn in the attempt at philosophically grasping what scientific theories and evidence suggest to us about the fundamental characters of our world, and at helping to find a new research strategy.

---

[1] Tanzella-Nitti 2009, p. 3.

[2] McGrath 2001-2003, v.1, pp. 45-50.

[3] According to Lonergan (1957, p. 417) "a heuristic notion [...] is the notion of an unknown content, and it is determined by anticipating the type of act through which the unknown would become known".

3.1 The Irreducibility of Happenings

3.1.1 *Is Everything Reducible to Laws?*

We have seen that both science and philosophy have the ambition of universality. However, when this aspiration becomes uncritical, it may give rise to a sort of "absolutism of reason". Philosophy, more than once in its long history, showed a certain devaluation of empirical data and even claimed that, by the sole means of reason, it is possible to settle any matter a-priori, and therefore tried to build itself as a *universal system*. Some tendencies to take this direction were already at work in the Middle Ages, and modern science could even be understood (or at least it was understood by its fathers) as a reaction to such an overweight of reason[4]. The most radical expression of this philosophical attitude, however, may be found in the early Modern Ages, especially in Cartesian philosophy with its dream of a *mathesis universalis*[5] (Subsec. 2.1.2) and in Spinoza's *more geometrico* system[6], as well as in Hegel's idealistic philosophy. In the latter case, we have the most radical expression of an overestimation of reason. It is well known that Hegel affirmed that "what is real is rational as well as what is rational is real"[7]. Since, however, a total resolution of what we call reality in rationality seems impossible, Hegel stated that the work of thinking is to unveil the unfoundedness of the world dissolving its apparent reality whenever it shows to be incongruent with the requirements of reason[8].

Such a hubris is not only typical of a certain philosophical attitude. Many scientists have also often forgotten the empirical methodology underlying modern science and have tried, as mentioned, to elevate certain disciplines or scientific standpoints to an all-encompassing rational world-view (Subsecs. 2.1.6 and 2.2.3). Indeed, the well-known Laplace's dictum[9] states that, if one could have complete knowledge of the present state of the universe in its totality (which is already an astonishing supposition in itself), one would be able to predict any future state of the world thanks to the laws of classical mechanics. This demands a *faith* in the absoluteness of classical-mechanical laws since it presupposes that: 1) our theories

---

[4] Bacon 1620; Galilei 1612.

[5] Descartes 1637; 1641.

[6] Spinoza 1677.

[7] Hegel 1821, Introduction; 1817, §6.

[8] Hegel 1817, §50; 1821, § 324.

[9] Laplace 1820.

perfectly mirror reality, so that the laws we formulate are *literally* "the laws of nature", and 2) we exclude, even hypothetically, unexpected facts but assume mechanical laws unconditionally[10]. Another interesting example is represented by those like Lord Kelvin who, near the end of the 19th century, seemed to assume that physics was essentially accomplished as a scientific discipline[11]: indeed, in his 1900 address to the British Association for the Advancement of Sciences, he stated that "there is nothing new to be discovered in physics now. All that remains is more and more precise measurements". This sounds really astonishing if we think about the developments of quantum mechanics and relativity theories few years later. Here, we find again the striving for the establishment of a closed theoretical system. Moreover, as we have mentioned (Subsec. 2.1.6), one of the main recent trends in physics is to try to reach a so-called Great Unification or "Theory of Everything". As to this, the issue is to properly understand what is actually meant by this expression. Some may think that with the achievement of such a theory, science (physics, to be precise) will be able to explain any single fundamental character of our universe (since any force and physical law can be understood as being generated by a process of a stepwise break of symmetry) and predict all the main properties of bodies and forces. This seems to be unwarranted and is not even compatible with the actual status of specialization characterizing scientific activity (Subsecs 2.2.4-2.2.5). Finally, we emphasize that these kinds of generalizations seem to be out of the reach of any experimental trial[12].

The problem is even deeper. Throughout the examination of our first proposed principle, we shall see that several bodies of evidence, coming from different scientific domains, suggest a certain irreducibility of natural happenings to universal laws. This may be seen as a manifestation of the contingency of our world, which must be regarded as the fundamental character of any elementary *event*. Therefore, laws do not abrogate such a character, given that they can be regarded as ruling uniform and regular processes occurring in a more general framework in which even contingencies occur[13].

In the following subsection we shall examine the form of inference called induc-

---

[10] This is why A. Comte (1830-1842, II, Lect. XIX) said that at that time astronomy was the only discpline to have fully and rigorously liberated itself from any – direct or even indirect – theological or metaphysical influence.

[11] Alexander 2001, p. 231. See also McGrath 2005, p. 103.

[12] Feynman 1988.

[13] Pannenberg 2008, pp. 31-32.

tion, after our presentation of deduction and abduction in Subsec. 2.1.3. The reader may wonder why we treat induction in this context and deduction and abduction in the previous chapter. The reason is that the treatment of deduction and abduction was very relevant for dealing with the issue of the foundations of scientific theories in a proper way, whereas the proper context for analysing induction is the irreducibility of happenings, which have triggered the historical change of paradigm brought about by the birth of quantum mechanics.

3.1.2 *The Birth of Quantum Mechanics and Induction*

Throughout this section, we shall show how relevant empirical data are for the furthering of science, and that recent scientific theories stemming from striking experimental and theoretical findings suggest more and more that happenings in the natural world are neither fully reducible to regularities or laws, nor entirely predictable in their singularity and variability. Let us start this examination from the most basic and empirically tested physical theory of our era: quantum mechanics[14]. This revolutionary theory stemmed from a series of totally unexpected findings, in a period when the feeling that physics was almost complete was widespread, as mentioned above. As recalled (Subsec. 2.1.3), the path-breaker of this new development was Max Planck[15], who showed how the blackbody problem may be solved by assuming that electromagnetic radiation (light) behaves discontinuously when interacting with matter. In 1905, drawing from Planck's assumption, Einstein was able to provide an explanation for the photoelectric effect (i.e. the emission of electrons from a thin metal sheet stroke by light).[16] The core of this problem, quite synthetically, is that, if one takes light as a purely undulatory phenomenon, the emission of electrons from the metal sheet should take far more time than that experimentally observed. If, on the contrary, light assumes a corpuscular behaviour, the almost instantaneous emission of electrons may be easily accounted for (in this case, the photons, that is, the light quanta, can be conceived as sort of bullets impinging upon electrons and scattering them outside the metal surface). Another important evidence yielding the birth of quantum

---

[14] For an historical reconstruction, see Kuhn 1978. For a technical account, see Auletta *et al.* 2009a, Ch 1.

[15] Planck 1900a, 1900b.

[16] Einstein 1905.

mechanics was the discovery of the Compton effect[17]. The effect occurs when X-rays (another form of electromagnetic radiation) are scattered after collisions with electrons, and it is observed that the wavelength of the scattered ray is slightly longer than that of incident rays, implying that it has less energy than the latter. This means that a part of the energy and momentum of the incoming photons is transferred to the electrons. The crucial point is that measuring the Compton wavelength of the electron, when its rest mass and the speed of light are known, precisely yields the Planck constant value. This was one of the pieces of evidence that convinced the majority of physicists of the physical (and not only formal) relevance of Planck's quantization principle.

The quantization principle was also fundamental for dealing with the problem of the stability of the hydrogen atom. Rutherford's classical planetary model of the atom was indeed affected by the problem that a negatively charged particle (the electron) orbiting around a positively charged one (the proton) should rapidly fall into the latter (due to the electromagnetic force inducing an attraction between the two particles), thus making the stability of atoms that we usually observe impossible. N. Bohr[18] found a solution to this problem by further extending Planck's principle, i.e. by hypothesizing the existence of *discrete* permissible stable stationary orbits for electrons around the nucleus, each specified by a characteristic energy value: in order for an electron to jump from one orbital level to another it needs a discrete amount of energy represented by the emission or absorption of photons. This point is also related to another fact grounding quantum mechanics: the discontinuity observed in the spectrum of the electromagnetic radiation emitted by diluted gases, which can be explained by maintaining, in accordance with the previous results, that electrons only orbit with specific frequencies around nuclei.

Now, all of these bodies of evidence came about in a largely independent way from one another and, more importantly for us, they were originally dealt with without modifying the theoretical framework of classical mechanics. Indeed, Planck's first conjecture was simply meant to point out that the black-body problem could be solved by using the laws of classical electromagnetism with the proviso that in certain particular circumstances, i.e. when interacting with the granular structure of matter, light assumed a different behaviour (we have called such an inference abduction). Also in the case of Bohr's solution to the problem of the sta-

---

[17] Compton 1923.
[18] Bohr 1913.

bility of atoms, although introducing a restriction on the allowed stationary orbits of electrons around nuclei, *within* these orbits, it was assumed that the behaviour of the electron was still ruled by the laws of classical mechanics.

It was only the *accumulation* of this sparse and partially *independent* experimental evidence (showing that nature manifests many radical discontinuities where, according to classical physics, we should find continuities) that led the physicists' community to the conviction that partial corrections to the classical world-view were insufficient, and that the formulation of a *new*, alternative theory was necessary.

The process summarized here is what we call *induction*, and it comes into play any time a previous theoretical system is about to be *substituted* by a new theory. In order to do this, one single piece of evidence or even some striking but sparse evidence could not suffice. In general, indeed, scientists faced with single local problems try to amend the theory with specific corrections – by means of *abduction* (Subsec. 2.1.3). The very accumulation of disparate and independent evidence is precisely what is needed to dismiss a theoretical framework, and this is the essence of induction[19]. It is well known that a similar process occurred when the new Copernican view of the world substituted the old Ptolemaic system[20]. However, as for abduction, we need an *insight* in order to fulfil this inferential process (which in itself is inconclusive or only negatively conclusive and thus can at most lead to the realization that a new generalization is needed[21]). As previously stated, the only difference is that, in the case of induction, the issue at stake is to give rise to a new theoretical framework, and not to correct a previously existing one. In the transition from classical to quantum mechanics, Heisenberg[22] had the insight that led to the foundations of the new theory. He assumed that the physical parameters describing a quantum system no longer commute as ordinary numbers do but need to be expressed in terms of mathematical operators[23].

---

[19] Auletta 2009a.

[20] Galilei 1632. See also Kuhn 1957.

[21] What is generally called induction, that is, the inference of some regularity from the observation of some cases, should rather deserve the name of *extrapolation*. Unfortunately, it is of less use in science and philosophy since it can at most lead to the inference of some statistical regularities (e.g. that half of Italians possess a second house) while in those disciplines we need to deal with laws, their consequences, their application range and their validity.

[22] Heisenberg 1925.

[23] We say that two numbers a and b commute when we have $ab = ba$.

It is worth mentioning that induction, deduction and abduction constitute a circle[24]: induction is the form of inference opening the path to the formulations of new theoretical constructs; deduction tries to find out what the consequences of the assumed hypotheses may be, abduction tries to rectify the body of these assumptions and to explore anomalous domains of application. In this way, theories, although finally grasping ideas of general validity (constituting the basic intelligibility of our universe), are in themselves dynamical constructions based on inferences and being characterized by all the systemic stages of any other process of our world: formation, consolidation and repair, and decline[25] (Subsecs. 2.1.4-2.1.5).

We may consider the so far summarized theory of inferences as one of the major accomplishments of philosophy in a long and controversial history lasting many centuries, whose beginnings can be found in Plato's dialectic and Aristotle's[26] theory of syllogisms and whose climax is represented by Peirce's work[27]. Therefore, this examination can be considered as an aswer to the second question raised in Subsec. 2.2.7 about the admissible forms of inferences. At the basis of this general theory of inference there is the idea that in each inferential process there are three basic statements[28]: one expressing a law or a rule, one indicating the set of objects to which that rule applies, and finally a result that one expects by applying that rule to this set of objects. Expressed in such general terms, an inference can be taken to be a deduction. However, since this expectation may be contradicted by events occurring in subsequent experience, we face the alternative of rejecting either the criterion for having chosen the set of objects or the rule. In the former case, we perform an abduction, in the latter an induction. Therefore, deduction is by its very essence *anterograde* (see also Subsec. 2.2.2), because it starts with assumptions and tries to understand what the possible experiences could be as a consequence. On the contrary, both abduction and induction are *retrograde*, because they start from experiences that have contradicted expectations and go back to the foundations of knowledge in order to see what assumptions need to be rejected. In this way, although all the three forms of inference are necessary in their inferential form, abduction and induction, which lead to a negative or inconclusive conclusion (i.e.

---

[24] Peirce 1901, pp. 96-97.

[25] Odum–Odum 1976, pp. 62-63.

[26] Aristotle *An. Pr.-An. Post.*

[27] Peirce 1866; 1869; 1878a; 1901; 1903.

[28] Auletta 2009a; 2011d and literature therein.

to the need of formulating new properties or new laws), have a result that is indeterminate[29]. In the case of deduction we cannot speak of such an indetermination but still of a certain uncertainty affecting future or possible experiences. This is the reason why deduction improves our understanding of certain laws or principles.

3.1.3 *Quantum Events*

Let us now consider the main novelties that were introduced by quantum mechanics. The most striking feature of quantum-mechanical systems was to display a *genuine random component* within the most basic level of physical reality. This is what provoked the astonishment in classical-minded scientists since it could be taken to be a violation of the order assumed to be present in nature (Subsec. 2.3.5). This problem is very complex and needs to be assessed not only in this section but also in the next ones. For the time being, let us clarify a crucial point: quantum mechanical *laws* are fully deterministic, as classical laws are, but they do *not* rule the singular occurrence of happenings (such as, typically, detection events).[30] On the contrary, they rule deterministically the time-evolution of *probability distributions* of those happenings, whilst any particular event cannot, in the general case, be predicted with certainty, and hence it should be regarded as being genuinely random. For instance, we can predict on average how long it takes for a piece of radioactive matter to reduce to half, but we cannot predict *which* atom will decay next. To take another simple example, suppose that we prepare photons polarized along a certain direction, say 45° with respect to the vertical direction; this means that the systems are all *exactly in the same state*. Suppose also that we want to let such systems pass through a filter for vertical polarization. Quantum mechanics predicts that of all the systems undergoing the test, exactly half will pass through the filter (the rest being absorbed by the filter itself), which is bizarre from a classical point of view as systems in the same state should behave

---

[29] It is Aristotle himself to apparently draw this conclusion, since he says (*An. Post.*, 90b5-7) that all the inferences of the second figure (that, following Peirce, we have associated with abduction) are negative (they therefore represent a not-positive inference of a property) and those of the third figure (that, again following Peirce, we have associated with induction) are all particular (they represent a not-conclusive inference of a law).

[30] As perfectly understood by Peirce (1891, p. 296) long before the birth of quantum mechanics, and in a time dominated by the deterministic world-view. See also Cartwright 1999, Ch. 7. In other words, quantum laws cannot be understood as statistical regularities.

in the same way. Moreover, there is *no way* to predict which individual photon will pass the test and which will not. In this specific sense, single quantum events are irreducible to regularities. Of course, they are not absolutely unbound from regularities when *constrained* by a series of factors inherent to both the involved quantum systems and their mutual relations in determined environmental and contextual conditions; however, such constraints *cannot determine* a single happening[31]. Obviously, this historical turn in the way physics considers reality was not easily accepted by the community of physicsts[32]. We have already mentioned Einstein's reaction against this violation of deterministic causality and also Bell's historical contribution in formulating a mathematical expression that was able to provide a test for this issue (Subsec. 2.2.6). Between the EPR paper and the formulation of Bell's theorem, D. Bohm transformed Einstein's initial insight in a detailed physical explanation of the classical type of how quantum events are produced[33]. According to Bohm, there are variables hidden from the current status of research (called therefore *hidden variables*) that are able to give rise to causal interconnections of the classical type between quantum events that apparently are not causally related. This interpretation implied the existence of a special quantum potential able to drive quantum particles and so to explain their weird and unpredictable behaviour (as in the mentioned example of photon polarization). A classically deterministic hidden-variable theory was disproved by the subsequent experimental tests of the Bell theorem[34]. Moreover, the assumption of a quantum potential led finally to the formulation of an implausible theory showing non-local features that were potentially clashing with special relativity[35]. Therefore, quantum events are to our knowledge really irreducible to any causal explanation.

---

[31] Auletta *et al.* 2009a, Ch 1.

[32] It is worth mentioning that in 1960's Filippo Selvaggi (professor of philosophy of nature at the Pontifical Gregorian University, the heritage of the *Collegio Romano*) already understood the fundamental relevance of quantum mechanics for upgrading the philosophy of nature (see Selvaggi 1964, Chs. 7-8).

[33] Bohm 1952; 1953.

[34] Auletta *et al.* 2009, Ch. 16.

[35] Bohm-Hiley 1987; 1989; 1993; Bohm *et al.* 1987; see also Auletta 2000, Ch. 32. It is interesting to note that a broad class of reasonable realist but non-local theories has also been experimentally disproved by Gröblacher *et al.* 2007.

Is this irreducibility of happenings a new result brought about by quantum mechanics? Yes and no. The negative part of the answer is due to the fact that the mentioned findings unveil a fundamental tenet that characterizes modern science from the beginning: the ultimacy of empirical facts. Subsequent developments have in part obscured this awareness by assuming that theories and laws can provide a total account of empirical facts (Subsec. 2.3.3). We may recall here the examples of Laplace and of other scholars in the last two centuries (Subsec. 3.1.1). The positive part of the answer relies on the fact that quantum mechanics provides a hint towards a *general* explanation of this circumstance, pointing out that laws and theories do not rule single events.

It is therefore quite surprising that a famous scientist like S. Hawking first presents himself as a follower of Laplace, assuming a principle of scientific determinism, according to which, given the state of the universe at any time, a complete set of laws fully determines both the future and the past, so that there is no mind, no free will, no miracles, and no necessity for a God, and then admits that quantum mechanics allows laws to rule only probabilities so that there is no single past or history of the universe[36]. We find this procedure to be quite astonishing and even contradictory. At the very least there is some confusion here, because if the word "scientific determinism" only means that natural phenomena should be explained in naturalistic terms (Subsec. 2.2.5) and admitting that many events and phenomena are not subjected to a classical determinism but rather to a determinism of probabilities, then the reference to Laplace and classical determinism is inappropriate. If understood in the other way, then we cannot fully grasp how it should be in accordance with the epistemology of the model-dependent realism proposed by the authors (Subsec. 2.1.2). Is determinism outside of the range of this epistemology? If so, determinism would became matter of faith. Even from a scientific point of view, to say that something like the human behaviour could be explained in pure physical terms (although we ordinarily do not because "we cannot solve exactly the equations governing the gravitational interactions of every atom in a person's body with every atom in the earth"[37]) appears to be a little bombastic if we

---

[36] Hawking-Mlodinow 2010, pp. 29-34, 72 and 82. This is why E. Mach (1905, pp. 282-86) already warned us that there is no possibility of deciding once and for all in the dispute between determism and indeterminism. This was written in 1905. There are now additional empirical reasons for suggesting that nature rather presents many situations in which indeterminism holds.

[37] Hawking-Mlodinow 2010, p. 32.

consider that we cannot solve the equations (but need to use approximations) even when only three different bodies (even macroscopic) are involved (as it is the case for a *single* helium atom).

3.1.4 *Complex Systems*

It may be possible to circumvent the above conclusion by assuming that the general features (randomness, unpredictability, and the peculiar character of laws) shown by quantum systems are confined only to this class of natural phenomena (or at least to the effects directly derived from them). This would be, however, quite naïve and essentially wrong from a strictly physical point of view. Indeed quantum effects are never lost, even within macroscopic or at least mesoscopic systems[38]: they may sometimes turn out to be negligible, but to completely wash them out would require an *infinite amount* of time[39]. Moreover the irreducibility of happenings could also be concluded by starting from *complex systems*. A system is complex if its behaviour depends crucially on its own details[40]. This implies that it is also strongly sensitive to small random fluctuations, which in many ordinary classical systems are absorbed, whereas in complex systems can be very quickly amplified. Indeed, a complex system shows long-range correlations, some of them also having a feedback character[41]. As a consequence, for complex systems it is also impossible to perform predictions about single behaviours, but any prediction is probabilistic in its fundamental nature. Indeed, the fact that a system is complex (or chaotic) and its dynamics shows extreme sensibility to the initial conditions, implies that the knowledge of the initial state of the system does not allow for predictions about its state after a sufficiently large time-interval. How large such time-interval must be to lose the possibility of predicting its state depends on some characters of the physical system at hand and on the degree of exactitude with which we know its initial state. This seems to imply that the more accurate the knowledge of the initial situation is, the larger the time interval is within which we can predict the evolved state. However, the unpre-

---

[38] Brune *et al.* 1996 ; Friedman *et al.* 2000 ; Julsgaard *et al.* 2001.
[39] Paz *et al.* 1993; Auletta *et al.* 2009a, Ch 9.
[40] Parisi 2006.
[41] Ellis 2005; 2006.

dictability of the time-evolution of the complex system after a certain time-interval does not reflect a mere limitation of our technological devices, since it is impossible *in principle* to determine such initial conditions themselves with arbitrary exactitude, due to the combination of the two crucial aspects already mentioned: long-range correlations and random fluctuations, which are amplified quickly in most cases. We would also like to mention that for many classical physical systems that are not complex, there are also situations in which it is impossible to determine with infinite accurateness the value of a physical magnitude when it is represented by a real number, since this would imply the knowledge of an *infinite* number of decimal digits[42].

3.1.5 *Happenings in Biology*

The irreducibility of happenings may also be found in the biological domain. Evolutionary biology assumes that a chance component intervenes in the variability triggering the evolution of species via natural selection. Darwin himself introduced such a mechanism as an explanation of evolution while being aware of the fact that natural laws only express *general* regularities of nature, thus not determining any *particular* detail. Indeed, in his famous letter to Asa Gray of May the 22nd, 1860, he wrote about his inclination to look at everything as resulting from general laws, with the details, whether good or bad, left to the working out of what we may call chance[43].

Nowadays, we know how genetic mutations can be brought about by a number of physical or chemical factors. Even if our knowledge of the mechanisms generally involved in the origin of genetic mutations is growing more and more, given the margin of (quantum and complex) fluctuations involved in such numerous chemical reactions, molecular assembly, and recombination, one cannot fully predict *the specific mutations* that will be produced, also taking into account their (internal and external) complex environment. For instance, DNA-repair mechanisms are highly efficient but not infallible; errors in the molecular checking mechanisms

---

[42] Auletta 2011a, Subsec. 2.2.2.

[43] See Darwin *Corr.* The same ideas are formulated in a letter to J. D. Hooker of July the 12th, 1870. It is also worth mentioning that such statements were expressed by Darwin with a certain reference to theological issues. On this point, see also Zycinski 2006a, pp. 29-31.

are very likely to be random. Presently, we also know that, apart from the physical-chemical factors mentioned above, there are also a number of strictly biological factors inducing genetic variability, especially during mitosis and meiosis. Transposable elements are likely to be responsible for the major part of the genetic recombination occurring in the whole history of life on Earth[44]. Genetic recombination, crossing over, and gene conversion occur (at least in sexually reproducing organisms) also in order to increase genetic variability across generations[45]. For example, once a so called "double Holliday junction" (i.e. a four-stranded DNA molecule generated by the binding of two different double-stranded DNA molecules) is formed in homologous recombination events during meiosis, it can be resolved (i.e. disentangled) in two different ways, one of which gives rise to crossing-over, the other does not; which of the two resolutions occur is random.

Furthermore, the acknowledged crucial importance of *epigenetic processes* in evolution brings us to realize the actual difficulties in tracing the relations between genotype and phenotype. Again, even if our understanding of the mechanisms that are generally involved in the development of a mature phenotype from the embryo are growingly investigated, one cannot fully predict the *punctual morphology* characterizing organisms in a natural context, also given the indirect yet still effective influence of environmental fluctuations since the very early stages of development. Epigenetic processes display the importance of newness in the biological domain much more than the early mechanisms identified by Darwinism[46]. It is likely that the emphasis on newness coming from the knowledge of epigenetic processes will cast more and more light on this issue also in other domains of life sciences.

3.1.6 *Brain Development and Mental Events*

Epigenetic processes play a crucial role in the development of the animal and human brain. The gross topological anatomy of the brain is largely determined by the genetic program, as it can be ascertained even by the elementary consideration

---

[44] Mahillon-Chandler 1998; Auletta 2011a, Sec. 9.7.1.

[45] Alberts *et al.* 2008, 304-316.

[46] Schlichting and Pigliucci 1998; West-Eberhard 2003; Wagner 2005; See also Pannenberg 2008, p. 92.

that it is shared by all the individuals of the same species[47]. However, such anatomical organization is by no means able, by itself, to give rise to a properly functioning brain. As a matter of fact, differences have been recorded even in genetically identical twins at the level of the finer topological organization of the cerebral cortex[48]. At the neuronal and synaptic level, differences become more significant, even in the case of isogenic individuals, especially in the visual system[49].

Brain development occurs in three different stages[50]:

(1) Neuron proliferation, death and migration;
(2) Aggregation, specification and transient connection among neurons;
(3) Synaptic strengthening and selection.

Although the first two stages proceed in an ordered way regulated by the species-specific genetic program, the last one, mainly occurring from birth to puberty[51], is strongly affected by the activity of the neurons and depends on individual experience and inputs coming from an interaction with the environment. The "Theory of Neuronal Group Selection"[52] strongly stresses the importance of *selection* during brain development (through neuronal death, elimination of collateral branches of neurons, and elimination of synapses in surviving neurons). Such a selection follows the *complex* interactions of the individual with the environment and the *unpredictable* responses of neurons to such stimuli. It is particularly important to mention the process of *re-entry* through which multiple vertical and horizontal neural interconnections are established especially when information that has already been processed in higher brain areas feedbacks into primary sensory areas[53].

Interesting examples of irreducible happenings are also represented by mental events, especially when quick conscious decisions are made[54]. It is a common fact that we are forced to switch from an unconscious modality of action to a conscious

---

[47] Kandel *et al.* 2000; Changeux 2009, pp. 152-153 and 196.
[48] Hasnain *et al.* 1998; Changeux 2009, pp. 185-186
[49] Macagno *et al.* 1973; Levinthal *et al.* 1976 ; Changeux 2009, p. 187-188.
[50] See Auletta 2011a, Subsec. 11.5.2.
[51] Changeux 2002, p. 190.
[52] Edelman 1987; 1992. Changeux–Danchin 1976.
[53] Edelmann–Tononi 2000. Auletta 2011a, Subsec. 11.5.3.
[54] Auletta 2011a, Sec. 21.3.

one, when unexpected events surprise us[55]. For instance, when driving a car, especially along a known path, we may do it in a fully unconscious way (indeed, we are often simultaneously mentally engaged with totally different kinds of problems). However, when there is danger (e.g. another car crossing our path), we become suddenly aware. It is interesting to note that, immediately after, we switch again to an unconscious modality of action, giving rise to a sequence of acts (like braking or turning) that could not be executed under conscious control, since the latter is more time-expensive and thus ineffective at avoiding the danger. This suggests that we have evidence of consciousness when there is a sudden jump from one unconscious context (the habitual driving) to another one (the exceptional sequence of acts to avoid the danger). The actions performed in one unconscious context are in general fully unrelated to those performed in the other one (to turn suddenly is not a consequence of driving mechanically along a known path). This is a consequence of the modularization of the brain, of which the habitual driving and appropriate reactions against dangerous situations are examples. Now, the relevant point is that the brain, when we are conscious, is a complex system showing long-ranging correlations among distant neurons and areas (as brain-imaging shows). However, the mental *event* of becoming suddenly aware is an irreducible happening, whose effectiveness consists precisely in the ability (thanks to those long-ranging correlations) to associate specific but very distant neural processes or brain areas. When this connection results in a new finding that can be considered to be interesting (i.e. grasping a new solution to a problem), there is the occurrence of what we call *insight*[56] (Subsecs. 2.1.3 and 3.1.2). Therefore, insight, as a mental event, can be defined as the novel joining of two contexts that are distant from a brain information-processing point of view.

3.1.7 *Epistemological Considerations*

From an *epistemological* perspective, we may wonder whether or not this irreducible variety and unpredictability of happenings constitute an insurmountable obstacle for scientific progress. A good epistemological principle is the following: Everything must be presumed to be explicable till proved inexplicable[57]. In fact,

---

[55] Lonergan 1957, p. 207.

[56] Lonergan 1957, p. 28.

[57] Peirce 1866, p. 399.

without the variability and the irreducibility of facts and events in our theoretical constructions, science would no longer receive the necessary stimuli to correct and change its theories and might remain trapped in a rigid theoretical system (Subsec. 3.1.3), thus losing its character as an effective knowledge that has a universal aspiration but is still anchored in *specific* fields and problems (Subsec. 3.1.1). As a matter of fact, science continues to achieve remarkable results, despite the limits of its theories. We come back here to the fundamental difference between assent to facts (happenings) and inference or hypothetical reasoning[58] (Subsec. 2.2.3). Obvioulsy, this assent does not need to be solely about facts proving our theories. When we are confronted with negative results (i.e. our hypotheses are refuted) during our scientific investigations, we have learnt something (Subsec. 2.1.5) and are able to make further consistent advances in our knowledge along two possible paths: either by abductively updating our theories and laws showing that they have unexpected domains of application (Subsec. 2.1.3), or by inductively abandoning old theories in favour of new theoretical constructs (Subsec. 3.1.2). Of course, we also learn about the world when the answer is positive, i.e. in the case where our theories "work", or our experiments do not disprove our constructs. Interestingly, in this case we probably advance less in acquiring new knowledge, but are obviously more confident about previously formulated hypotheses, so that these confirmations are a natural part of the consolidation process of our theories (Subsecs. 2.1.4 and 2.3.5). We shall turn in Sec. 3.2 to the astonishing *accordance* that our theories have with the world, which somehow represents the other side of the coin.

In conclusion, a natural world characterized by occurrences that are not fully predictable by any theoretical constructs, nor reducible to them, *guarantees an endless research for science,* instead of hindering or discouraging it[59]. The irreducibility of occurrences "only" implies that natural phenomena should be considered as the standard for evaluating our theories, as assumed at the very beginning of modern science, and not the other way around, which implies that theories and laws can only be assumed conditionally. This permits a continuous approximation towards a full understanding of natural reality that, nevertheless, remains *in a horizon of intelligibility* (and truth) never fully grasped by our theories[60] (Subsec. 2.2.8), which remain fallible and perfectible, and thus always open in their addressing that horizon.

---

[58] Newman 1870, Chs. 2-4.

[59] Peirce 1898b, p. 48.

[60] Lonergan 1957, p. 197; Maréchal 1926, pp. 217-236.

3.1.8 *Ontological Import of Irreducible Happenings*

From an *ontological* perspective, the irreducibility of occurrences has very relevant scientific, philosophical and theological consequences:

1. It provides the necessary condition for the *emergence*[61] of true novelties in nature throughout the history of our universe, presenting typical features not reducible to the underlying level of reality they emerged from: chemistry cannot be reduced to physics, nor biology to chemistry, and so on.
2. It hints at the role played by chance and random fluctuations in the very constitution of the natural world, especially for what concerns the relations with the organized complex structures that we have to deal with.
3. It grounds the *contingency* of our world (the 8th question raised in Subsec. 2.2.7), that is, the fundamental limitedness, locality, individual distinctiveness and irreducibility (including irreducibility to environmental conditions) of the finite being[62].
4. It also grounds the possibility of freedom, by making our world not intrinsically determined by inescapable laws of nature that would undermine the ability to choose.

Philosophers have indeed shown a certain sensitivity to the issue of individuality of events and entities. We may recall the work of Aristotle in which he singled out the notion of substance from all of the categories[63], emphasizing that it cannot be taken as a category in the proper sense of the word. Indeed, he used the Greek demonstrative τόδε τι to denote the irreducible individuality of the substance[64]. We may also recall the deep reflection of the Middle-Age philosopher and theologian Duns Scot on the irreducible *thisness* or *haecceitas* (deriving from the Latin demonstrative *haec*) of the individual[65]. Also the early modern philosopher Leibniz pointed out that in our universe there are no individuals that are perfectly identical. He formulated this assumption as a principle of the identity of indiscernibles, according to which two individuals could be identical only if they shared all prop-

---

[61] Chaisson 2001; Morowitz 2002; Murphy–Stoeger 2007
[62] Torrance 1985b. Pannenberg 1993, p. 98.
[63] Aristotle, *Phys*, 185a 31-32.
[64] Aristotle *Cat.*, 3b 10-23.
[65] Duns Scot *Ord.* II, d. 3. See also Peirce 1896; 1806.

erties. To show that this is impossible it suffices to take into account their different location in space and time[66]. However, the first philosopher to have thought about the possibility to formulate a general principle of nature centred on emerging variety and impredictability of events was C. S. Peirce[67]. Therefore, on the basis of the most recent results of science and in accordance with Peirce's philosophical standpoint, the *first fundamental principle* that we hypothetically propose as the groundwork for a renewed philosophy of nature is:

> A fundamental dimension of reality is represeneted by the *spontaneity* of happenings grounding the irreducible *variety* of nature that we observe.

We points out here that this is a general character of reality that ultimately can only be experienced and not conceptually articulated as such, and for this reason expressed at best through the work of poets and artists. In this way, we answer negatively to Einstein's radical question about chance (Subsec. 2.3.5). Summing up, admitting an ontological irreducibility of happenings in nature shows the inconsistency of understanding natural processes as ruled by a deterministic necessity (i.e. the common assumption that having certain initial conditions, singular effects will follow with necessity). This result is the critical import of the first principle and can even be understood to a certain extent as a sort of *"no-go theorem"* (therefore, in a first approximation, can be seen as an answer to the first philosophical question raised in Subsec. 2.2.7). The point is really crucial since the question naturally arises: Is any necessity of nature banned forever? Can the world be reduced to random fluctuations only? We are perfectly aware that, by introducing an *ontological irreducibility* of happenings to regularities, we are also postulating a "hampering" gap between the two. In order to deal with such a momentous problem, we are forced to turn to another order of considerations and to take into account the second aspect involved when dealing with the relations between our theories and nature: their astonishing agreement. Subsequently, we shall try to find a reasonable connection between these two aspects.

---

[66] Leibniz 1686b, § 9; 1689b, pp. 519-520.

[67] Peirce 1892a; *CP* 1.25, 1.302, 1.357, 2.85, 2.89. Peirce also called this principle *tychism*, meaning absolute chance and freedom (see *CP* 6.322 and also 1892b, p. 312). See also Auletta 2011a, Ch. 2 and Subsec. 6.5.1, 10.2.4-10.2.5, Secs 11.2 and 25.1.

3.2 Why Theories Accord with Nature

If our world consisted only of irreducible happenings, it would not be an ordered *cosmos* and, as a consequence, it would not be intelligible at all. As a matter of fact, primary experiences of happenings are not connected with any insight[68]: any original experience of an unexpected event has not yet been framed in an interpretative horizon, and in its uniqueness ultimately has the value of a negative feedback (something disproving a previous expectation). This is why we have said that any inquiry starts with doubts raised by problematic facts (Subsecs. 2.1.3-2.1.4). In the words of Peirce, genuine doubts have always got an external origin, usually out of surprise, from unexpected happenings[69]. We think that the awareness of this circumstance is fundamental in the empiricist tradition, especially when considering Hume's theory of impressions[70]. On the other hand, we are able to find out *laws* and *theories* that "work" throughout the world, which would be impossible unless nature actually possessed *stable* regularities and characters on its own, making it understandable; this is what *intelligibility* is all about.

3.2.1 Quantum Probabilities and Features

In the previous examination of the irreducibility of happenings, we started from the present most basic theory, quantum mechanics, which doubtlessly attests to such a character of our world in a radical manner. Now, the very astonishing fact is that quantum mechanics allows us to calculate the *probabilities* to obtain certain events with an incredible exactitude (for instance, to predict with which probability we would obtain certain measurement outcomes). Therefore, the question is: if the reality described by quantum mechanics only consisted of random events, how could we possibly formulate such probabilistic predictions? Looking at the structure of quantum theory, we shall realize that the Schrödinger equation (ruling the dynamics of any quantum system) shows a perfectly deterministic evolution of the so-called probability amplitudes, that is, the quantities allowing us to com-

---

[68] Lonergan 1957, p. 50.

[69] Peirce 1905b, pp. 348; 1868. This is why T. Reid (1764, Ch. 1, Sec. 3 and Ch. 2, Sec. 5) did not take Descartes' doubts seriously as far as they are posed by the mind that seeks for its own foundations and do not arise from certain fundamental experiences.

[70] Hume 1748, Secs. 2 and 5. See also Spencer 1860-1862, pp. 123-132; von Helmholtz 1867, pp. 610-611.

pute probabilities of possible outcomes. Now, when the state of a quantum system changes and therefore the probabilities of obtaining certain events also change, a prediction about the latter would be strictly impossible if the probability amplitudes computable at any moment were independent of each other. This is not the case because, in the state of any quantum system, *constraining correlations* relating all probability amplitudes are somehow nested, thus allowing us to formulate deterministic predictions about all outcome probabilities. In a recent paper, some of us[71] have called such interdependences quantum "*features*", for two reasons:

(1) They are those *characters* of the system that, not being properties in themselves, nevertheless determine the probabilities. Indeed, properties are local by definition, while features, being interconnections among components of the system, are non-local by their very nature.
(2) The term *feature* is sufficiently general to cover a wide range of quantum phenomena. In most of the scientific literature, entanglement (a concept already introduced above as a solution of the EPR problem: Subsec. 2.2.6) is considered to be the paradigmatic form of interdependence among quantum systems (e.g. two particles) not involving any exchange of signals. However, what is less understood is that entanglement is only a multi-particle *manifestation* of a most basic reality (which we call *features*) that is still present even in an *isolated* quantum system.

Given the latter proviso, entanglement is also a good example for dealing with the general issue of correlations. If we consider two entangled particles in a so called "singlet state", they manifest an amazing feature. If we measure their intrinsic magnetic momentum (called *spin*), we shall find out that, if one particle is in a spin-up state along a direction of an external magnetic field, the other one is necessarily in a spin-down state in the same direction, or vice-versa. That is, we shall find out *either* down-up *or* up-down, but *never* up-up or down-down. The statistics in which we would obtain all four cases (up-up, down-down, up-down and down-up) with equiprobability would be completely *random*. This is precisely the kind of statistics that we should expect if the world consisted only of happenings. To give a simple example drawn from our daily life, let us suppose that two people repeatedly throw two coins. If the number of tosses is sufficiently large, we would

---

[71] Auletta–Torcal 2011.

114

expect that on average each person would get *tails* half the amount of times, and *heads* half the amount of times (this is an example of the law of great numbers). If subsequently, the two people compared their tosses, they likely would find that the joint distributions of the two series of tosses is totally random, that is, we have all four possibilities (tail-tail, head-head, head-tail, tail-head) occurring without any specific order. If, on the contrary, they were to find that only tail-head and head-tail pairs occur, this would be really amazing and would search for an explanation. This is precisely what happens with entangled quantum systems.

The crucial point here is that to get a subset of possibilities out of a random set represents an increase in order. Indeed, if we get a tail, we are able to predict that our partner will get a head and vice versa. Translated in quantum mechanical terms, when we measure a quantum system and get spin-up as an outcome, we are sure that a possible partner measuring the twin particle will get a spin-down. Therefore, the general lesson we are trying to draw here is that the existence of quantum interdependences, that is, at a more general level, of *constraints*, is precisely the piece of the world that allows us to make *predictions* and to formulate theories and laws which have an *ontological* import (Subsec. 2.1.1). Hence, the irreducibility of quantum events discussed above (Subsec. 3.1.3) does not prevent us from attributing an aspect of regularity and lawfulness to the quantum world that is somehow opposite to the irreducibility of happenings but may nonentheless be articulated together with the latter[72]. The price to pay here is that these laws no longer have the deterministic character attributed to them by scientists like Laplace. Better said: they still have a deterministic character but do not rule single events, as said in the previous section, but probabilities and probability amplitudes. In other words, the feature that enables quantum mechanics to make predictions is precisely the same one that limits the applicability of its laws. This can be seen as a second no-go theorem (see Subsec. 3.1.8).

Finally, we emphasize that features are forms of interdependency without necessarily implying interaction or an exchange of signals, as Schrödinger understood for the first time when he introduced the concept of quantum entanglement (again Subsec. 2.2.6). As we shall see, this insight allows the understanding of the canalization of processes occurring at a lower level of complexity without any direct intervention[73] (Subsec. 2.2.5).

---

[72] See also Tanzella-Nitti 1997, p. 18-20.
[73] Auletta *et al.* 2008.

3.2.2 *From Quantum to Matter: Convergences*

Now, it may well be that things stand in this way solely in quantum mechanics, so that we are not allowed to apply the same conclusion to other aspects of reality. Yet, it turns out that the above general considerations also apply to other kinds of systems. Remarkably, in our universe correlations are not formed once and for all. When matter structures like atoms and molecules emerge, quantum features like entanglement play a less relevant role, because localized properties connected with physical quantities like mass, energy and electric charge become much more relevant. Nevertheless, we have here again the rising of different astonishing interdependences, which find an aesthetic expression in the geometric shape of many molecules. It is well known that F. A. Kekulé guessed the ring-like chemical structure of benzene by dreaming of a snake seizing its own tail, a beautiful example of insight (where the distant contexts joined are represented here by the snake and the chemical problem, Subsec. 3.1.6) as well as of the application of quantum mechanics to molecular objects since all of the electrons of the ring are shared. In other words, benzene is an ordered configuration from both a quantum-mechanical and a geometrical-molecular point of view.

The existence of correlations helps us to understand a further general and surprising fact of our universe: the existence of *convergences* at any scale. To take an example, let us consider that to have a hydrogen atom, the simplest chemical element in our universe, many constraints must be satisfied. We shall not bother the reader with minute technical details but simply touch upon some fundamental issues. Abstractly speaking, to put together two particles – a proton and an electron – seems to be very easy. However, there are several requirements to be fulfilled. The most reasonable one is that the involved particles must be stable, otherwise we would not have a stable atom either. It is interesting to observe that the proton is a particle composed of quarks. Now, most of the quarks' combinations are short living or unstable, and this strongly reduces the number of possible candidates. Moreover, the fact that the proton is composed of quarks, and is therefore heavier than the electron (which is an elementary particle, as far as we know) is relevant, since the proton (1840 times more massive than the electron) occupies the nucleus of the atom and therefore represents the centre of stability of the whole atomic system – an issue which is even more relevant when there are many protons and electrons, as is the case in heavier elements. Another remarkable aspect is that the proton and the electron have electric charge (positive and negative, respectively). The two charges need to balance (which excludes many abstract combinations of quarks that would not give the appropriate charge value). Moreover, a dipolar force like the electromagnetic one is much more suitable than a mono-polar (like the gravitational) one when we need a combinatory for producing the different chemical elements. Indeed, to take the example of the solar system, the Sun attracts any possible body independently of its distance and mass (even if, obviously, there are constraints restricting the range of possible distances and other involved factors). On the contrary, in the case of a dipolar force, we must have the same number of differently charged particles, and this more easily permits a combinatory (when we do not have such a balance as for ions, this contributes even more to the establishment of connections). Therefore, to put two particles together in order to build the simplest atom of the universe is a sophisticated process of convergence in which many physical parameters must necessarily fit[74]. This aspect is, of course, even more important when heavier elements are formed. Indeed, the combinatory of elementary particles at the bases of the variety of elements constituting our universe and our planet allowed for the theoretical prediction of the existence of some of such elements. It would be fruitful to understand the constitution of the hydrogen atom and of the atomic structure of the universe in general as the result of a *selection process* through which many alternative possibilities are discarded (or have never arisen at all). It appears, moreover, that the stability of matter we usually observe and therefore its ordered configuration, instead of being somehow hindered by the peculiar characters of quantum systems, including their irreducible randomness, is rather *allowed* by them through the process of convergent stabilization that we have just sketched[75].

3.2.3 *Long-range Interrelations in Complex Systems*

When we turn to complex systems, quantum interdependencies are again not so relevant, but new and remarkable correlations are enabled. In particular, we have pointed out that also for complex systems we cannot formulate predictions about single behaviours, since correlations determine a quick spread and amplification of even small fluctuations (Subsec. 3.1.4). The relevant point, however,

---

[74] Thirring 2007. Thus, we find it to be astonishing and even childish to declare that there can be universes where electrons have "the weight of golf balls and the force of gravity is stronger than that of magnetism" (Hawking-Mlodinow 2010, p. 142). This seems to be deprived of any scientific basis and can be considered a consequence of the so called multi-verse understanding of physics.

[75] See also Barrow 1990, p. 197.

is that we can formulate general predictions here about *classes* of behaviours: again this would be strictly impossible if there were no long-ranging correlations among parts and components of complex systems (correlations are, therefore, both the problem and its solution).[76] For instance, when we heat a liquid confined between two metal layers, above a certain temperature threshold the so-called Bénard cells are formed. These are hexagonal structures that arise due to liquid motion and can be either levogyre (turn left) or dextrogyre (turn right). They must spatially alternate, that is, if one cell is, say, levogyre, it is surrounded by dextrogyre ones. The amazing fact is that the whole global, coordinated configuration arises suddenly, without the possibility that a cell somehow communicates its rotation direction to the neighbourhood ones. Therefore, the structure must arise due to global correlations without signal sending and *not* as a sequence of local interactions, in full accordance with what we have said previsously about quantum features.

The general conclusion of this brief examination is that, as the complexity of the systems grows, certain correlations may well be disrupted, although the capability of giving rise to new correlations is maintained. The establishment of new correlations allows new and growing levels of complexity that would otherwise not be reached. Therefore, nature challenges us to find out an explanation of this fact at a higher level of generality. We would like to express such a generalization as a *second fundamental principle*:

Nature displays the capability to exert *constraints* and to *canalize* phenomena giving rise to convergences at all levels of complexity.

It is worth mentioning that there is a millenary history in philosophy and theology with reagrds to the concept of relation or interdependency. In the ancient world, relation was generally understood as being accidental. Indeed, it was considered in this way in Aristotle's treatise *De Categoriae*[77]. It is likely that the first scholar to have understood the ontological relevance of relations was St Augustine when he dealt with the difficult issue of the relations between the three Persons of the Holy Trinity[78]. However, the first scholar to have conceived the possibility of relations without

---

[76] Parisi 1999.
[77] Aristotle *Cat.*
[78] Augustine 399-422.

interactions was G. W. Leibniz when he famously spoke of *monads* as being interconnected without communicating with each other[79], an amazing speculation at that time. Leibniz also assumed that the universe consists of a huge web of interconnections, referring to a universal harmony: σύμφονια πάντα[80]. This point of view was extended by C. Peirce, who distinguished between pure correlations (second-order or static relations in his terminology) and interactions (third-order or dynamic relations).[81] Peirce also considered the relation meant as "constraint" to be the core of a general principle of nature that he called *secondness*[82]. Therefore, this principle is well established in the history of philosophy. However, it is worth emphasizing that it was physics to have determined the destiny of this concept, in a early stage through the classical-mechanical rejection of the ontological import of relation (as supported by Leibniz) whose clearest expression is the separability principle of EPR (Subsec. 2.2.6), and at a later stage by providing empirical evidence for the existence of interdependences introducing the concept of quantum entanglement. This very short reminder shows again not only how fruitful philosophical and theological investigations can be for science in anticipating important hypotheses about the world but also the circumstance that philosophy is able (when receiving empirical support by science; Subsec. 2.2.3) to arrive at results that can be considered as permanent and very valuable acquisitions of the human mind (Subsecs. 2.2.7-2.2.8).

---

[79] Leibniz 1712-1714, §§ 7, 57 and 62.

[80] Leibniz 1712-1714, § 61.

[81] Peirce, *CP* 1.293; 1.303-332; 3.472-473.

[82] Peirce said (*CP*, 1.325): "The idea of second is predominant in the ideas of causation and of statical force. For cause and effect are two; and statical forces always occur between pairs. Constraint is a Secondness". In the same paragraph he stressed that this is the way in which, in a psychological context, "the past appears to act directly upon the future". This means that a constraining influence cannot be taken as a dynamical action. Also elsewhere (Peirce 1887-1888, p. 171) he said: "The genuine second suffers and yet resists, like dead matter, whose existence consists in its inertia." Peirce also called this principle *synechism*, intending it as a principle of continuity (Peirce 1892b). It must also be added that Peirce is not always coherent and sometimes seemed to confuse such a passive constraint with a form of active action or reaction. For instance, criticizing Hegel for not having taken into account secondness, he said (Peirce 1887-1888, p. 179): "He has committed the trifling oversight of forgetting that there is a real world with real actions and reactions" (See also Peirce *CP*, 5.469; 1888, p. 211). First-order relation is absence of relation, that is atomic happening.

3.2.4 *Information and the Formal Dimension of Nature*

There is a natural question that now arises: what is the kind of reality that the correlations in our universe consist of? We indeed live in a world that is physical, but correlations are by definition *formal* (Subsecs. 2.2.5-2.2.6). The problem is that correlations and physical quantities are tightly connected: it suffices to say that correlations are instantiated in physical systems and enter into interactions involving exchanges of physical magnitudes. Now, how is it possible to put together something formal with something else that is not? We need a sort of quantity that is both formal and nevertheless linked to the material dimension. Let us come back to the example of entanglement that we have proposed above (Subsec. 3.2.1). We have shown that, in an entangled state, the outcome statistics are more ordered (two out of four possible cases) than when there is no entanglement (four out of four cases). There is a language for dealing in the most general way with such kinds of problems: the language of *information*. Indeed, information (and its connected quantity, the so-called Shannon entropy[83]):

(a) Is concerned with the issue of singling out a subset of elements (the message or the information we like to acquire) from a larger set (the set of all possible messages or at least the set of elementary units out of which any message can be composed, like the alphabet);

(b) Measures the unlikeliness of performing such an extrapolation, which is expressed by the probability that the latter does not occur or is not chosen. Obviously, if there is no (syntactical) order among the units and the sequence of the latter is random, to acquire or to guess the right message will be much more difficult than in the case in which there is some rule (for instance, in some cases, we can understand that an encrypted message represents an English sentence because of the higher or lower frequency of some letters).

Therefore, we may say that information and entropy are concerned with the amount of *order* and disorder of a system. The more correlations there are in a system, the more the system is to be considered ordered and the easier the guess is about its state. Indeed, the way in which entanglement can be mathematically described is with the so-called *mutual information*, that is, the information shared by several systems or components of a system. The total amount of disorder (i.e.

---

[83] Shannon 1948.

the Shannon entropy) of a compound system, whose components are subsystems a and b, is equal to the amount of disorder of a, taken separately, plus the amount of disorder of b, taken separately, minus their *mutual information*. In other words, when there is no correlation at all, i.e. when the mutual information is zero, the total amount of disorder is maximal, being simply the sum of the disorders of the two subsystems taken separately (in this case the two systems are totally disconnected).

Since the sum of the disorders of the two subsystems taken separately represents indeed the maximal amount of entropy that is attainable, when these two systems are put together randomly, it is in fact possible that some interdependencies are established by chance, that is, some amount of order spontaneously arises. This makes a very common experience understandable, e.g. when we find some shape or regularity in objects like clouds or constellations. Obviously, the established shape may be completely arbitrary or metaphoric, but the order that has been individuated may be not. So, this circumstance, which has been used throughout history by many philosophers (and scientists) as an argument for showing the fragility or even subjectivity of human knowledge, tells us on the contrary how some order can arise from a less ordered initial situation[84]. This does not imply a violation of the second law of thermodynamics (stating that the entropy of an isolated system either grows or remains constant) because, when putting two systems together, the *maximal entropy attainable* is bigger than that attainable by the separated systems[85], since the number of configurations that are possible with the compound system is higher than the sum of the numbers of configurations of the subsystems taken separately. For instance, if we have three systems each of which can be in two states, the number of the possible configurations of the three systems taken separately is six, but the number of possible combinations when putting the three systems together is eight[86]. Another consequence is that maximal entropy is a rather limiting case and that, in any part of our universe, a certain amount of order is always present (ultimately because some correlations are always established).

---

[84] "Any plurality of objects whatever have some character in common (no matter how insignificant)" (Peirce CP 6.402).

[85] For a technical account see Auletta 2011a, Chs 2 and 6.

[86] The number would be even higher if all possible permutations were considered, that is, when the order of the sequence matters, as it is ordinarily the case when we deal with messages exchanging.

Information is something much more basic and widespread than it has been thus far considered by the scientific community[87]. Actually, quantum systems can be considered as information codifiers and even as information processors[88]; indeed, the most advanced frontier of quantum mechanics is the so-called quantum computation[89]. Recently, matter structures, space and gravitation have been considered as emerging from a basic level of information nested in our universe at the microscopic level[90].

Then, the question naturally arises: how far can we make use of the formalism of quantum mehanics to solve other kinds of problems? Apparently, not in a generalized way, because e.g., it has been shown that we cannot make use of this physical theory for solving problems like free will or human consciousness[91], which brings us to the conclusion that the mind or the brain are not quantum-mechanical engines as tentatively assumed by R. Penrose[92]. Nevertheless, there is a general formal lesson that is precisely related to the role of information in nature. A physical theory that satisfies the requirement of causal connection set by the relativity theory (according to which there is no exchange of superluminal signals) needs to satisfy a certain bond on the possible interconnections between distant systems. Classical-mechanical theories ruling a part of our macroscopic world, in which only local interactions are acknowledged (according to the separability principle, Subsec. 2.2.6), are known to satisfy a much lower bond. Quantum mechanics is grounded on long-ranging correlations also in absence of local interactions and therefore violates the classical-mechanical bond but satisfies the relativistic bond. Now the question is: is any physical theory located between the relativistic (non-local) and quantum-mechanical bonds possible? Theoretically speaking, yes. However, this would conflict with the information causality principle that tells us that the amount of information that a receiver can recover is bounded by the quantity of information actually transmitted (in other words, this principle relates to the amount of information that an observer can gain about a set of data belonging to another observer[93]). Since this is impossible, we have as a consequence that quantum mechanics sets precise con-

---

[87] Wheeler 1990.

[88] Auletta 2011a, Ch 2.

[89] Nielsen–Chuang 2000.

[90] Verlinde 2011.

[91] Clarke 2010.

[92] Penrose 1989.

[93] Pawlowski *et al.* 2009. Auletta 2011c.

ditions on any information exchanging in our universe and therefore also on its causal interconnections. This shows that quantum information is not irrelevant when we deal with information exchanging at *any* level in our universe, also in biological systems or through the operations of the mind. Obviously, a lot of work still needs to be done in order to understand the specific ways in which these issues are related. Nevertheless, we may assume that some very basic features of quantum information are true of any system of our universe.

DNA may also be considered as an example of information codification. It is quite interesting to observe that, in this case, a sharp separation between chemical bonds and information combinatory is necessary. As we have said, information is a formal quantity. For instance, the sequence of words on a written page *cannot depend* on the chemistry of the page. Similarly in the case of DNA, the four bases (A, G, C, T), the basic elements of the code, are not involved in the chemical bonds constituting the sugar-phosphate backbone along a single strand, that is, the linkage of a nucleotide with another is not constrained by the chemical details of the bases. This is the reason why the bases may instantiate truly informational connections along the same strand, to ensure that any sequence is possible in principle and, therefore, that DNA is able to store information[94].

In conclusion, we stress that what is interesting with information is that it is a quantity "sitting" somehow between pure mathematics and physical reality (we recall here the 5th question raised in Subsec. 2.2.7). It must necessarily be instantiated in some physical media but, as we have seen, the informational content is not dependent on the specific physical characters of the medium (although the latter must satisfy some general requirements to carry information). The information carried by a physical medium rather resides in the structure, the order, the configuration, or the combination of discrete physical elements. We cannot avoid recalling here what Galilei, one of the fathers of the scientific revolution, expressed about the world, namely that it is written in mathematical characters[95]. This view is deeply rooted in Plato's philosophical approach (Sec. 1.2). The general lesson might be that our universe shows an intrinsic intelligibility and this is the reason why our theories accord with reality. Is this conclusion justified? We need to consider some further aspects to deal seriously with this problem.

---

[94] Auletta 2011a, Sec. 7.4.1.
[95] Galilei 1623. See also Tanzella-Nitti 2009, pp. 128 and 151.

3.2.5 *Potentiality*

Let us come back to the case of DNA. In itself, it represents latent information, i.e., information that is not being used for something else. As a matter of fact, it is not DNA but RNA that directly concurs to the building of a protein. In other words, DNA, when unexpressed, is something *potential* (potential information, indeed). To produce any effect at all, it needs other, external conditions represented by a cluster of proteins and RNAs that are able to give rise to the complex process of transcription, and only in this case is able to produce a string of instructions represented by the so-called messenger RNA.

Potentiality is an issue that is strictly connected with the formal dimension of our world. Constraints and structures, although being fundamental constituents of our world, are not active by themselves and need to be *activated* in order to give rise to some physical consequences. They can therefore be considered as potential relative to these consequences[96]. In general terms, potentiality should be understood as a necessary but not at all sufficient condition to give rise to certain effects. A useful and simple example is the conformation of the trees in a forest[97]. By itself, this structure is not able to produce any effect actively. But, when an efficient cause, like wind or fire, occurs, the structure is able to canalize the action of this factor thus contributing to the specific effect produced (e.g. to direct fire in a specific pathway or to block wind). We must avoid a possible confusion here. Potentiality does not mean lack of existence, or a diminished form of existence. Any structure or correlation is fully existent, and as far as we can understand, speaking about different degrees of existence is meaningless. The only relevant issue here is that a structure, by itself, is incapable of producing effects, since it is inactive and displays no functionality if it is not stimulated (activated) by something else, just like a guitar needs to be played to emit a sound.

Potentialities are widespread in nature, at all levels of complexity. It is not by chance that even in the framework of neo-positivism, the issue of "dispositional properties" arose[98]. In quantum mechanics they are easily recognizable in the so-called entangled states[99] but should be acknowledged (especially if related to the

---

[96] Aristotle *Phys.*, 201 a 10-19. Already Averroes, *AM*, p. 45 (j51) had already said that we predicate 'potencies' of dispositions and forms only because they are sometimes active and sometimes not.

[97] Auletta *et al.* 2008.

[98] Hempel 1953, § 6.

[99] Heisenberg 1958.

central notion of *potential information*) as one of the constitutive ontological characters of quantum systems in terms of features (Subsec. 3.2.1).[100] Other examples of potentialities are the different forms of fields or the Gibbs free energy expressing the potentiality of a chemical to do work. In biology, potentiality is manifest in DNA, as already discussed, and also in many bio-molecular patterns that become activated only in response to appropriate stimuli, as occurring during the epigenetic process. In the brain, it is known that many excitation patterns activate when faced with environmental cues or cognitive stimuli. Even memory represents a set of dispositions that needs to be activated within the context of an occurring perception or experience[101].

It is appropriate here to recall Aristotle's doctrine of matter and potentiality. The opinion that any potentiality comes from matter is quite common in philosophical literature. This is to a certain extent true if we consider that a material system shows dispositions or conditions for doing something additional or different relative to its current state, as it is clearly the case for biological systems. From this correct point, however, some philosophers assume that matter represents a sort of "privative" condition relatively to the implementation of some configuration or state of affair. This second statement is not true. We do not need to recall here that for Aristotle matter without form (the so-called primary matter) does not exist precisely for the reason that it would be a fully privative concept and therefore a kind of pure abstract possibility (without any specification) and not a kind of potentiality[102]. In this sense, it would be nothing rather than something. On the contrary, when we speak of potentiality, we are dealing with a reality endowed with certain faculties or dispositions to produce something else if properly acti-

---

[100] Auletta 2005; Auletta-Tarozzi 2004b; Auletta–Torcal 2011.

[101] Laroche *et al.* 1995. See also Auletta 2010, Ch. 17. This seems a point badly understood by Mill (1865, Chs. 8 and 10) and the associationist tradition in general (see Fabro 1941a Chs. 1-2).

[102] Aristotle is very clear about the fact that what is potential is always something very determined (*Met.*, 1047b35-108a2). See Averroes, *AM*, pp. 103(j106). On this point see also Selvaggi 1985, pp. 526-28. When Aristotle speaks of a substratum (*hypokeitai*), as for example in *Gen. Corr.*, 319b2-4, he distinguishes it from being (which should be informed matter), although it is not always clear what he precisely means, as explained in William's commentary in his edition of the treatise.

[103] Averroes, *AM*, p. 44 (j50-j51) says that the concept of potency is expressed in many ways, and in particular (i) we attribute potency to that which causes change in something else in so far as it causes change in something else, no matter whether such potencies are physical or ra-

vated[103], and we may further assume that these dispositions are due to formal elements or structures contained in material bodies. To take an example drawn from current research, when two quantum systems are entangled during a so-called EPR protocol, they share potential information that can be used later on to acquire further information that is classically inaccessible[104]. This shows that what we are dealing with here is something that displays a certain ability to produce some effects. Unfortunately, since many scholars connect *potentiality* to matter as such, which is in turn understood in a privative sense, there is sometimes a certain degree of devaluation attached to this notion in the aristotelic-thomistic tradition[105]. It is therefore worth mentioning that Aristotle himself clearly distinguishes between (i) form, (ii) matter, and (iii) absence of form or shortage (specific privation), by clearly denying that matter can be identified with the latter[106]. Obviously, we cannot repropose today Aristotle's conception of matter as such. However, our proposal does not seem to be in conflict at least with the spirit of Aristotle's *Physics*. It is also true that in many places Aristotle thinks of form as being actual and not potential[107]. It seems however opportune to distinguish here

---

tional, (ii) the term is predicated of potencies whose nature is to be set in motion by something else, (iii) is predicated of all that has in itself a principle of change, (iv) is also predicated of what is able to perform a good (appropriate) action, as well as (v) of all that is scarcely affected and not easily destroyed. It seems therefore that he was clear about the fact the potency cannot be disjointed from a certain capability to act or react in a certain way. See also Averroes, *AM*, pp. 68-69 (j73-j74). See Lewis 2009, p. 172.

[104] Horodecki *et al.* 2005; see also Auletta 2006b.

[105] Having established this, it can become a terminological problem of whether these resources that produce certain consequences should be called potentialities or active indeterminacy (Selvaggi 1964, pp. 151-52) considering that we agree that these resources are connected somehow with formal structures and constraints and are therefore not privative (they are not expressions of what Selvaggi calls passive indeterminacy). Obviously, the distinction between *potentia activa* and *potentia passiva* has a ground in scholasticism and can even be found in Aristotle, *Met.*, 1019a15- b15 (see also Hamilton 1859, I, Lect. X, pp. 175-178). Also Averroes, *AM*, pp. 95-96 (j98-j99) distinguishes between an active and passive potency but adds that the first is mostly due to a combination of nature and art while the latter is only due to art and therefore can only be activated by a human agent. Finally Cajetan, in his treatise *On God's Infinity* (1568, t. III, tract. II, p. 192) says that to any active potency corresponds "aliquid possibile ex potentia passiva" and excludes that there is a *potentia neutra* in natural things (1568, t. III, tract. III).

[106] Aristotle *Phys.*, 192a3-13. See Lewis 2009, p. 166.

[107] For instance, *Met.*, 1042b10-11 or 1043a29-31.

between what he called *energeia* (which is more dynamical) and what he called *entelecheia* (which is the final accomplished state, in which there is presumably no longer potentiality, at least from the point of view of the attainment of this state).[108] If the form is associated to the *energeia* (as Aristotle often does), then it does not mean a form that has been fully activated or realized from the start but something that is in the process of realization. Otherwise, we need to assume a kind of preformationism (the prexistence of a complete active or activated form from the start in any growth or generation process) that is in sharp contrast with Aristotle's epigenetic tenet[109]. Then, it would be better to say that what is potential is informed matter[110] (matter *with* form at a particular stage of some developmental or dynamical process) and that matter can be taken to be potential in relation to form only to express its role as a *causal factor*: matter represents the conditions of possibility of certain processes, i.e. the set of elements that make possible something else (like the variety of biomolecules allowing the existence of living organisms).

Thus, in which sense could formal constraints play a causal role and therefore have an explicative power? As we have stressed (Subsec. 2.1.1), any true explanation should rely on some exhibition of the causal mechamisms involved. However, we have also seen that quantum-mechanical processes do not show mechanical causation in the traditional sense of the word. What we suggest is that formal constraints contribute to the production of several events and effects without simulatenously possessing the dynamical power to bring those effects about by them-

---

[108] Blair 1992. A very enlightening text of Aristotle on this subject is *Met.* 1066a20-26.

[109] In the *Generation of Animals* (Book II, Ch. 2), Aristotle understood that there is a certain progression in the building of the organism and that "all the parts are first marked out in their outlines and acquire later on their colour and softness or hardness, exactly as if Nature were a painter producing a work of art, for painters, too, first sketch in the animal with lines and only after that put in the colours".

[110] It may be interesting to recall that Aristotle refuted to consider a third principle unifying matter and form as if they were two indiepndent entities, and so understood the substance as composed of the latter two as a whole (see Lewis 2009, pp. 175-76). From this point of view, Leibinz's attempt at unifying modern science with Aristotelism (see Introduction) suffers two main drawbacks: since he considerd the form only in terms of entelechy, he faced the problem of how to connect mind and body (the latter being rather an aggregate of monads), a problem that was posed by his correspondent des Bosses in terms of the *vinculum substantiale* and was only unsatisfactory solved (Leibniz 1706-1716); on the other hand he consequently supported a form of preformationism (see Subsec. 2.3.3).

selves. In this sense, they are explicative and can be called causes[111] (Aristotle called them formal causes) but do not represent exhaustive explanations. This is why one needs efficient, mechanical causes in order to activate constraints and make them effective[112]. *A fortiori*, this also implies that constraints cannot have any efficient role by themselves but can only exert a canalizing influence when properly activated. For instance, any epigenetic process can start only when appropriate environmental stimuli are present. In other words, a form, *when* activated, represents, from a *causal* point of view (as a formal cause, in Aristotelian parlance), the set of constraints that canalize a particular dynamical process, and therefore it cannot be said to be potential in the same sense in which we say it of a material cause (i.e. conditions of possibility). However, to hypostatize too much these two causal factors (material and formal causes) would imply a break of the unity of the substance as informed matter, which was fundamental for Aristotle. Unfortunately, one of the biggest misunderstandings especially of later Scholasticism and Modern-Age Aristotelism is to have assigned to the formal causes, and especially to the so-called substantial forms, an efficient causal power (for instance, producing certain accidents). This is likely due to a misleading attribution of a full actuality, and therefore an active causal power, to form. For this reason, the whole theory of formal causes was rejected by modern scientists as fully incomprehensible[113].

Finally, it is interesting to note that Whitehead was somehow aware of the different but complementary nature of potentiality and actual events: he said, indeed, that actuality is a decision amid potentiality[114] and is "incurably" atomic[115] (as expressed in events or happenings), whereas continuity concerns what is potential.

---

[111] We may therefore say with St Thomas that a cause is what brings a particular influx on another thing: see Selvaggi 1964 (pp. 55-60) for a quotation and examination of this point. Selvaggi also shows that modern neo-scholastics like Suárez have a different concept of cause that is more consonant with modern scientific determinism (see also pp. 143-45).

[112] It is universally known that for Aristotle a being cannot actualize its potentiality without something being already actual, in most cases even something else beforehand actual (e.g. *Met.*, 1049b17 and ff.). But also in the cases in which there is an intrinsic actuality, Aristotle would barely deny that this actualization cannot happen without an external stimulus, since the consequence would be again a kind of preformationism.

[113] As beautifully reconstructed in Pasnau 2004. This is why in Auletta 2008 there is a distinction between dynamical causes (efficient and teleonomic-teleologic ones) and non-dynamical ones (material causes as conditions of possibility and formal causes as constraints).

[114] Whitehead 1929, pp. 43-46.

[115] Whitehead 1929, p. 61.

3.2.6 Intelligibility

Correlations do not determine single happenings. As a matter of fact, when we have a quantum entanglement *both* joint outcomes up-down and down-up are possible. It is *only* when we get a local event (for instance, up) that we can predict the other joint event (in this case, down). This is why correlations are only potential and possess a general character as laws do. This is quite natural, because we have shown (Subsec. 3.2.1) that correlations are precisely the basis of any order in our universe and also of laws, and we have repeatedly said that laws only have a general character. Two consequences emerge from this consideration:

(1) Correlations constitute the texture of the universe's intelligibility, since they allow us to build predictions and to construct theories.
(2) The fact that correlations have a general character and do not determine single events is the way in which regularity turns out to be *compatible* with random happenings: it explains why events can be random as well as why occurring events can determine contexts that in themselves are only potential.

It is obvious that we still have to clarify how these two fundamental aspects of our world are positively related. By now, our main philosophical result, is that intelligibility itself concerns the overall regularities and structures of our world and not the specific ways or modalities through which those regularities and laws are implemented or displayed through specific local events.

From an *epistemological* viewpoint, intelligibility of nature is both a presupposition and a consequence of our cognitive enterprise.

- It is a *presupposition* to the extent to which, in furthering the inferential work characteristic of our enquiries about nature, we assume that the latter is indeed intelligible[116]. Only on this basis are we legitimated in believing that there must be further laws and formal constraints to which nature should obey. For instance, it is only because of such a fundamental belief that Galilei was able to infer that the astronomical observations of the raggedness of the moon could be interpreted as mountains and valleys (although the supposition that there were seas was obviously wrong). This inference was actually a guess. Indeed, it is interesting to recall that Feyerabend[117], fa-

---

[116] Lonergan 1957. See also Polkinghorne-Beale 2009.
[117] Feyerabend 1976.

mously a supporter of happenings but not so much of the intelligibility of nature, found such a step to be illegitimate. From a strict empirical point of view, he was right. The problem is that without these kinds of inferences, science is not able to progress (Sec. 2.1): we should have waited for a direct experience of the orography of the moon that has been done only recently through space missions. When we actually infer the existence of a huge black hole in the core of a galaxy by observing the gravitational distortion that it gives rise to, we are again performing an inference of Galilei's kind, those that Einstein[118] found to be constitutive for science. Moreover, as we shall argue in the following, the inferences that philosophy and theology of nature often perform are also of the same kind. Even if the formal object is obviously completely different, it might be said that the assumption of intelligibility is what natural sciences, philosophy of nature, and theology of nature share at a fundamental level[119]. In all these cases, the ground for ontological inferences is an appeal to intelligibility[120].

- It is a *consequence* to the extent to which, once we have inferred hypotheses and found that they accord with reality, we become more and more convinced that nature is *indeed* understandable. It is precisely because intelligibility is, at first, an assumption and not patent evidence or something that we could experience or somehow prove, we feel surprised and amazed in realizing that our particular theoretical elaborations have indeed got a counterpart in nature. It suffices to consider the enormous mathematical complexity of quantum theory and its incredible predictive power.

Although having answered in negative way to Einstein's question about chance in nature, we answer in a positive way to his concern about intelligibility in nature that represented the background of that question (Subsec. 2.3.5). However, he was unable to satisfactorily provide arguments supporting this assumption apart from the success of scientific theories. Then, Einstein's wonder about "the eternal mystery" of the comprehensibility of nature[121] and Wigner's puzzlement over the "unreasonable effectiveness of mathematics"[122] are well rooted but at the same time solved in their paradoxicality once we ascertain and rationally justify that fundamental character of nature that is its intelligibility: we can rationally ground our rationality only provided that we acknowl-

---

[118] Einstein *MW*, p. 159.

[119] Tanzella-Nitti 1997.

[120] Polkinghorne 2009, p. 24.

[121] Einstein 1918, p. 109; 1936, p. 61.

[122] Wigner 1960; See also Heller 2005.

edge that nature is also rational[123]. The reader may disagree on some of our conclusions but cannot deny the value of this question. However, we should also avoid incurring in Hegel's error (Subsec. 3.1.1) by establishing an isomorphism between nature and rationality. The reality of irreducible happenings hinders this possibility. However, the critical import of the second principle stated at the end of Subsec. 3.2.3 tells us that nature cannot be understood as a pure random collection of disjointed facts either. Therefore, a new more fundamental question emerges, as we shall see now.

3.3 The Dynamic Interplay between Regularities and Happenings

The two principles proposed above might be seen as presenting two opposite dimensions of nature. As mentioned, the first principle mainly concerns the substrate of what has usually been intended as "contingency" or "chance", whereas the second one, by indicating the relevance of regularities, clearly hints at the "necessity" characterizing the natural world. It remains problematic to explain how these two general aspects, both attested in the scientific enterprise, can be reconciled (see the 8th philosophical question of Subsec. 2.2.7). Most scientists and philosophers have felt the urge to accommodate these two different orders of reality in a common framework, although the elaborated solutions often ended up in emphasizing one of the two aspects as being the "fundamental" one. On the other hand, it is easy to see that the simple acknowledgment of the presence of both with no further critical considerations, will lead to a dualistic view, which in turn may leave the issue of the relations between the two principles in the real world at best controversially understood, and at worst outside of any rational enquiry. Hence, the difficult task is to *think* appropriately (philosophically) of the co-existence of happenings and regularities, or of spontaneity and order, without overlooking the ontological and scientific relevance of any of the two. To this purpose, we shall first take into consideration some basic features of living systems, particularly in their evolutionary dimension, for it is in the biological domain that we believe that suitable hints to a possible solution of this

---

[123] See Benedict XVI 2011. Also Averroes, *AM*, pp. 67-68 (j73) was sensible to this point, although he arrived at conclusions that we shall criticize below. Indeed, he says that there is nothing whereby a sensible thing could become potentially intelligible, that is, a potential object of thought, except that it originates from intellectual conceptualization. From this it follows that there are separate forms that exist which are the cause of the intelligibility of sensible substances although such substantial forms through which sensible things become potentially intelligible are given to sensible things only by means of nature and the celestial bodies, i.e. these forms are forms of the celestial bodies.

problem may be found. Then, we shall turn back to more general ontological and epistemological considerations in order to refine the proposed philosophical sketch.

3.3.1 *The Search for Evolutionary Stable Solutions*

The historical turn in biology represented by Darwin's formulation of evolutionary theory[124] is crucially related to the author's understanding that, across the generations, variation is continuously produced, and that the environment acts as a sort of selecting filter allowing only certain "solutions" to survive and to be inherited. It is an open question whether or not Darwin understood such variation as genuinely random. In the later formulation of this theory, especially in its neo-Darwinian form, as it can be found in the first half of the 20th century, a sort of classical-mechanical understanding dominated, according to which chance happenings could be essentially traced back to the fortuitous crossing of two independent causal chains: that of the organism and that of the environment[125]. With the subsequent development of genetics and the rise of molecular biology (in the second half of the 20th century), it became growingly evident that there are intrinsic random mutations in the genetic heritage of organisms (Subsec. 3.1.5). These genetic random mutations were regarded as being exposed to the action of the environment, representing a sort of necessary selective pressure impinging on biological species. Traditionally, not only the genetic transmission itself across the generations, but also the epigenetic building of a phenotype starting from a genetic program was seen in deterministic terms. A paradigmatic turning point is expressed in the famous work of Jacques Monod *Chance and Necessity*[126] in which these two concepts are conceived as follows: Monod thinks that life as well as biological species and new functionalities arise purely randomly (he expresses many times the idea that life should be considered as something that has a very low probability of occurring) while the subsequent natural-selection filter stabilizes certain solutions, rendering them necessary (bonding) for the descendants.

Monod's views are both innovative and also anchored in the knowledge of almost half a century ago. Let us start from the latter point. New results in biology suggest that[127]: 1) life solutions seem to be not so exceptional and indeed arose very early in

---

[124] Darwin 1859.

[125] Mayr 1988.

[126] Monod 1970.

[127] D'Ambrosio 2008.

132

the history of our planet; 2) the genetic program does not act so deterministically as it was understood at that time. In such a context, notions like teleonomy and teleology are becoming more and more acknowledged as important and distinct explanatory tools in biology[128]. Teleonomy refers to the organism's basic property of being endowed with a project for its own self-building. We emphasized that this is a concept already introduced by Monod himself[129]. This notion does not imply having goals and can be applied to the biological domain at the level of phylogeny, and also has interesting applications in epigeny. This notion can be widened so to understand the latter as a process in which the internal (genetic) program of the organism allows for starting a basic process through which environmental signals are used as cues for giving rise to structures and functions that are adaptive[130]. In other words, teleonomy in epigeny is a basic process of co-adaptation, and can also affect phylogenetic processes. When we like to fully understand co-adaptation, we need to consider also the ontogeny of individual organisms, especially when considering their ability to cope with variable conditions in their environment and to survive. It seems to us to be difficult to deal with this problem without employing the notions of goal and the control mechanisms that allow the realisation of this goal. This concept does not have an intentional import since goals may be implemented in the organisms also through their genetic and epigenetic programs. Accordingly, F. Ayala stressed that we need to distinguish between teleonomy and teleology in biology[131]. Although the notion of teleology cannot be applied to phylogeny or to the evolution of species, there is, nevertheless, a connection between ontogeny and phylogeny, as far as biological evolution is the result not only of the occurrence of genetic mutations (happenings), but also of the capability of organisms to integrate the occurred mutation in a complex network of control mechanisms that is able to canalize its possible effects or contributions to the overall metabolic system (representing here the regularity, i.e. the organismic stability). Moreover, different "solutions" to environmental pressures are possible, and the plasticity of phenotypes that is particularly high during development[132] ensures adaptive responses,

---

[128] See Ayala 1998b. See also Ayala 1970 and Auletta 2011a, Sec. 8.2.

[129] See Monod 1970, Ch. 1. The author understood such a notion in a narrower sense than we do here. See Ayala 1970, p. 9.

[130] Auletta *et al.* 2011a.

[131] Ayala 1998a-b, 2006 and 2007. F. Ayala was the first scholar to use both of these concepts as appropriate, indispensable, and non-reducible explanations in biology.

[132] Auletta 2011a, Sec. 10.2.

involving for instance, differential gene expression[133]. Finally, mature phenotypes display the ability to carve out the environment into a niche suitable for their needs; this ontogenetic activity will in turn have its relevant consequences on the subsequent generations, since the changed environment will again exert a constraining influence on the offsprings. The sketched circular process corresponds to a widespread co-adaptation (as a result of both teleonomic and teleological mechanisms),[134] in which certain biological novelties, although spontaneously arisen, are maintained thanks to particular constraints, and different specific constraints are slowly activated *as a consequence* of the diffusion and proliferation of those novelties. Therefore, we may say that it is *by virtue* of these constraints (and not in spite of them) that we have the diversification of life forms that we currently observe. These aspects may be summarized with the notion of *emergent evolution*[135].

The constraints we are speaking about are both of the external (environmental) and internal (organismic) kind. This is extremely relevant as the very interplay between these two kinds of constraints constitutes the co-adaptation process that we have just mentioned. For this reason, we are not fully satisfied with the approach recently undertaken by J. Fodor and M. Piattelli-Palmarini[136], which seems to consider epigeny as a sort of generalization of the genetic program but also sharing with the latter the impermeability to external influences, thus overlooking the contribution of environmental constraints and their involvement in co-adaptation processes. It should also be noted that the authors seem to dismiss too easily the concept of biological function and attribute no causal value to natural selection. We shall turn back to the issue of biological functions in the next subsection. With regards to natural selection, it is obvious to us that it cannot be detached from both (i) the source of variability and the constraints arising endogenously in the organism and (ii) the complex network of interrelations that a certain population establishes with other species in a certain environmental context. To isolate natural selection from these two aspects and elevating it to the sole explanation of phylogenetic phenomena (as it is often the case in traditional neo-Darwinism) presents the risk of reducing natural selection to an empty word. Nevertheless, when taking all of these aspects seriously into account, the fact remains that natural selection is the causal filter stabilizing certain solutions and can even be considered as the interface between endogenous and exogenous

---

[133] Gilbert 2008.

[134] Spencer 1855, pp. 385-388; Lloyd Morgan 1891, p. 19; Auletta 2011a, Ch 8.

[135] Lloyd Morgan 1923.

[136] Fodor and Piattelli-Palmarini 2010.

processes. In conclusion, we are not inclined to consider natural selection as insignificant or even to dismiss it as a sort of anthropomorphic explanation[137].

The general lesson we would like to draw from all that precedes is that the kind of necessity that we may find in nature does not consist in conditions acting in a mechanical way from the start but is rather the *result* of a dynamical process in which mutations and random events are *integrated* in a complex network of constraints, whose activation enables a growing *canalization* (Subsec. 3.2.5) such that the final result may display a certain necessity. Only this kind of necessity is compatible with a world in which there are both random happenings and constraints (as we have stressed in the introduction to this section)[138]. Moreover, we claim that this understanding of necessity is fully compatible with what we have epistemologically stated about scientific theories building their own foundations (Subsec. 2.1.2).

Let us consider a very relevant example, namely the evolutionary passage from prokaryotes to eukaryotes. This passage should not be understood as occurring due to some necessity from the outset, i.e. to some deterministic development at a chemical or molecular level, but should be considered from the point of view of the requirements that *become* necessary given certain previous developments: when the complexity of cells grows, it also becomes increasingly important to *modularize* certain functions (like DNA replication and DNA expression, transcription and translation, etc.), so that organisms becoming more and more complex are naturally canalized along the path leading to eukaryotes (essentially a teleonomic process). Obviously, in this final solution, there are many aspects that may well be contingent and result purely from chance; but there are likely many others (like DNA-packaging, the segregation of DNA within a nucleus, the metabolism confined in the extra-nucleic cytoplasm, and so on) that have a general value and can even be considered as necessary at that stage of biological complexity; otherwise the organism could not survive. This is indeed a selective fork. We see again how natural selection helps and is integrated into the emergence of new biological phenomena. This is the kind of necessity that we are speaking about: the result of a *convergent dynamical process*. We may assume that, starting from a pool of different cells, many will converge to this or similar so-

---

[137] As the mentioned authors seem to do in the note 7, chapter 5 of their book.

[138] It is worth mentioning here that Pannenberg (1993, pp. 105-108) develops some interesting considerations that are consonant with our own approach to necessity. Indeed, Pannenberg suggests that the regularities that natural science describes as laws originally emerge as forms out of dynamical processes.

lutions, evidence for which is provided by the fact that this whole process probably had a symbiotic character[139]. It may be also helpful to reconsider here the example of DNA previously introduced (Subsec. 3.2.4): we may say that the chemical structure of DNA is obviously the result of a contingent evolution and therefore of many random events that may and may not have actually happened. Nevertheless, the general characters of information codification, and especially the need to separate sharply between chemical bonds and information combinatory, is a *necessity* to which a family of sufficiently complex structures trying to couple a metabolic system with a genetic program will converge, due again to a selective fork.

Summing up, we should distinguish between the *selective pressure* on the organisms that can canalize their phylogenetic or ontogenetic processes, on the one hand, and on the other the ability of the organism to establish correlations with the environment and *activate constraints* within itself in the form of organized and ordered structures (Subsec. 3.2.3).

3.3.2 Biological Convergences

Convergences, already introduced with respect to the physical domain (Subsec. 3.2.2), are therefore a crucial feature of biological systems in their evolution, assuming here a distinctive dimension. The common view is that we have convergent evolution when similar structures and traits emerge in cladogenetically distant species. These characters are conventionally called "analogous" as far as they do not share a direct common ancestry (like the "homologues" do). Very famous examples of analogous characters due to convergent evolution are the camera eye in vertebrates and in cephalopods, or the forelimbs of bats being analogous to bird wings. It is true that some of those phenomena could be explained as "latent" homologies in different and even phylogenetically distant species, that is, in terms of a common unexpressed DNA root. However, examples are currently multiplying in scientific literature, and convergences are being acknowledged at all levels of the biological domain, from phenotypic morphology and physiology up to biochemical pathways. If true convergences at the level of DNA sequence are rare even though they are increasingly recorded, structural convergences in protein folding are sometimes amazing. What is particularly interesting is the example of two DNA-binding proteins separated by more than one billion years of evolution

---

[139] Margulis 1970.

(yeast α2 protein and *Drosophila* engrailed protein) having an almost identical three-dimensional structure and the same function (both are regulatory proteins of the homeodomain family) but sharing *only* 17 over 60 amino acid residues[140]; much of the latter, moreover, are not placed in the active domains. It is difficult to explain the incredible array of these examples by resorting to a pure genetic explanation without taking seriously into account the co-adaptation processes previously introduced; in such cases even parallel expression of homologous genes always requires appropriate conditions in order to emerge and to be stabilized in a mature phenotype, like specific environmental cues or appropriate regulatory networks[141]. Consequently, if we admit that convergences are ubiquitous in evolution[142], we need to provide a general explanation rooted in the distinctive characters and capabilities of all organisms.

It is true that living beings can exist in a potentially infinite number of different forms. However, in spite of the unimaginable number of DNA-base combinations and the overwhelming phenotypic variability that might occur given a certain genome (the so-called norm of reaction[143]), the number of organisms that can exist in a relatively stable way are actually quite limited. Moreover, it is also useful to consider that, as noted by Wagner[144], the more a clade diversifies, the more convergences are frequent, thus suggesting that the "space" of possible diversification is not unconstrained and inexhaustible. Now, the widespread convergence towards a limited number of stable forms of life is, in our view, the consequence of the two general constraints that all living beings are subjected to in relation to biological *functionalities*, namely: (1) there are general and specific functions at the different levels of complexity that any living being must fulfil to survive; (2) functions require more or less specific core-structural elements and highly recurrent architectures for their implementation (although we recall that there is no one-to-one mapping between – molecular or anatomical – structures and functions). Thus, the continuous production of variability that Darwin understood very well is a kind of necessary condition presupposed from the start leading to the emergence of different structures from the molecular level to the macro-phenotypic one when appropriate constraints and teleonomic-

---

[140] Wolberg *et al.* 1991.

[141] De Beer 1971; Wilkins 2002; Auletta 2011a, Sec. 9.5 and 11.7.

[142] Conway Morris 2006; 2008.

[143] Gould–Lewontin 1979; Gupta–Lewontin 1982.

[144] Wagner 2000.

teleological mechanisms are in place. Indeed, the crucial point is that, at the evolutionary time scale, from this pool of variability only structures that are at least potentially able to respond to the basic functional constraints are preserved.

3.3.3 *The Principle of Dynamicity*

As a matter of fact, biology and especially the theory of evolution has attracted the attention of the scientific world upon the crucial relevance of the dynamical processes of integration, whose most astonishing expression is the emergence of novelties and convergence on stable solutions[145]. Since the issue of convergences goes much further than those shown in the biological domain only (Subsec. 3.2.2), the question arises as to whether this dynamism is typical of biological systems only or constitutes a general character of our universe. In order to ascertain if the latter is the case, let us consider the deep significance of this dynamism as it is displayed in biology. One may wonder why, in order to address the general question, we go back again to the particular issue of biological systems. The reason is that those systems may be seen as a laboratory for studying such a problem, and it is here that we should find features that also have general validity, if any.

We have seen that the sources of novelty in biological systems are represented either by endogenous mutations or by external fluctuations and variations in the environment, inducing a perturbation in the homeostasis of the organism. Let us consider in particular the second aspect since it is easier to deal with; however, the reader should be aware that the first aspect is not in principle different, so that our treatment of the problem may also be applied to that case, thus being provided with sufficient generality. When there is a perturbation, organisms are driven to new ontogenetic and/or phylogenetic equilibria. Let us remark that without any *perturbing* happening in a given system in equilibrium, there would be no dynamics at all, but simply a preservation of its state. Reciprocally, if the system did not instantiate a web of interrelations and constraints on its own, and if it were not suitably co-adapted to its environment, such a perturbation, if sufficiently strong, would again give rise to no dynamics at all, but would lead to a sudden breaking-down of the system. If not sufficiently strong, the organism would preserve its state. On the contrary, the very fact that any organism instantiates a network of

---

[145] In Hawking-Mlodinow 2010 (pp. 67-68) there seems to be a flavour of the concept of emergence when it is said that single neurons do not tell us how brain excitation patterns happen, and neither do water molecules explain the behaviour of a lake.

constraints determines a situation such that, when this network is perturbed or endangered by an unpredicted perturbation, the organism is naturally pushed to counteract, in order to lower the effects of the perturbation as much as possible. This cannot be accomplished in a completely satisfactory way, otherwise the perturbation would not have arisen at all, since the system would have displayed a total control on its environment in advance. Therefore, the solution is *necessarily dynamic*, that is, *a provisional and open trade-off* that can neither wash out the perturbation, nor come back to the previous stable state.

It is very important to stress that the specificity of organisms may be found in their extraordinary ability to connect informational aspects with thermodynamic-entropic ones[146]. From an ontogenetic point of view, any environmental fluctuation is perceived by the organism as a mismatch between what it "expects" and what it actually acquires as an informational input. This mismatch is called *surprisal* in information theory. Therefore, in order to keep its homeostasis the organism needs to lower this surprisal as much as possible. Now, the crucial result is that lowering the surprisal corresponds to lowering the statistical equivalent of the so-called free-energy function[147] (which is what we expect in accordance with thermodynamics since in such a context it means an increase in entropy). In other words, this process is thermodynamically favoured, and consequently the preservation of the homeostasis is favoured too.

Since the latter considerations only apply to organisms, we may again ask if there are aspects that have a general value. What we consider as the general lesson is the following: every natural system once perturbed never goes back exactly to its previous state. We know very well that classical mechanics has taken for granted that when a system is perturbed, it will go back to its initial state if the perturbation is relatively small; otherwise, the system is driven to another steady state. However, this should be taken rather as an idealization that can often be quite unhelpful when dealing with real scientific problems. The great scientist Henri Poincaré developed the idea that there is always a cycle through which a system (as well as the whole universe) can come back to its initial state. Careful analysis showed that this cycle could be exceedingly long (even many times the age of the universe). Further studies[148] have led to the consideration that no system can come back *ex-*

---

[146] Schrödinger 1944.

[147] Friston and Stephan 2007; Friston and Kiebel 2009; Friston *et al.* 2010. See also Auletta 2011a, Chs. 7-8.

[148] Crutchfield *et al.* 1986.

actly to the same initial state, and that there will always be some irregularities and fluctuations. As a matter of fact, complex systems when slightly perturbed are often driven very far away from their previous state (Subsec. 3.1.4). They show, together with chaotic systems, what is technically called dynamic *itinerancy*, that is, the fact that across time they shift along a dynamic trajectory passing through many relatively unstable states, in the same way as organisms do along their ontogenetic path. The issue, however, is much more general. Thermodinamical systems are ruled by the so-called Le Chatelier-Braun principle, which states that any perturbation of a factor contributing to equilibrium induces a compensating change in an opposing factor[149] that results again in a dynamical shift. Even physical systems as simple as quantum-mechanical ones show a similar behaviour. The so called theory of the geometric phase[150] tells us that after a cycle, a quantum-mechanical system goes back to a state that shows a non-vanishing angle with the initial one (we remind the reader that quantum states are represented by vectors that rotate during the time-evolution of the system), and thus they show a noticeable difference.

Therefore, we have sufficient grounds to assume that this behaviour is a general feature of nature. Now, if we ask what the reason is for such a behaviour, thanks to the previous examination of biological systems we can provide the following answer[151]: the *dynamic interplay* between (random) happenings (i.e. perturbations) and constraints determines an integration process such that systems are continuously driven to further relative stable or unstable states. It is the basis for understanding in a new way the concept of system as a transient complex of interrelations (see the 5th question of Subsec. 2.2.7). Therefore, we propose as the third fundamental principle of philosophy of nature the following:

Nature displays an itinerant dynamic interplay between happenings and constraints.

This principle is characterized by a critical aspect (similarly for what we have seen for the first and second heuristic principles): to a certain extent it can be conceived as a principle of irreversibility. It is again a sort of "no-go theorem". In fact, our universe, in its totality, shows an irreversible process of complexification along

---

[149] Ulanowicz 1986.
[150] Auletta *et al.* 2009a, Sec. 13.8.
[151] Auletta 2011a, Subsec. 8.2.7.

its history. As far as we know, particles have emerged from a much more undetermined state, molecules emerged from particles, life from prebiotic chemistry, and more complex forms of life from more elementary ones.

Let us stress the circumstance that a research strategy cannot work without no-go theorems, which have been introduced here as consequences of the three heuristic principles. They are indeed the way in which scholars delimit the boundaries of their research and therefore the kind of methodology they will apply. The Platonic stress on science (in opposition to mere opinion) as dealing with eternal and perfect entities like those of geometry or astronomy[152] is precisely a no-go theorem that defines the character and scope of ancient science (Sec. 1.2).[153] The Modern-Age rejection of essences can be understood to a certain extent as opposite to the latter but its core is again a no-go theorem: the scope of science will be mechanical explanations only. It may incidentally be the case here that a certain explanation could turn out to be a beautiful mathematical result that a Platonist would consider to be an expression of the ancient research strategy, as it happens when symmetries are found. These are however rather limiting cases and not the way in which modern science is actually practised. We may think that Kepler's work represents the dividing line between these two approaches: after having looked for a Platonic explanation of heavens (based on perfect solids),[154] he finally had to admit that the planetary orbits were elliptic, which represented for him a violation of the Platonic research syntax[155]. Now, if we are about to find out a new research strategy (Sec. 1.3) we need again some clear no-go theorem. We think that to select explanations centered on a dynamics of transient attractors and equilbria is precisely the kind of no-go theorem that we look for (an attractor is a set encompassing all states of a relative equilibrium to which a system spontaneously tends). We also remark that the other two no-go theorems that we have found in Subsecs. 3.1.8 (about the impossibility to predict singular happenings) and 3.2.1 (about the general and probabilistic character of our laws that still allows us to predict classes of behaviours as pointed out in Subsec. 3.2.3), are consequences of the kind of dynamics we are envisaging here and therefore the latter no go-theorem includes the first two. This seems a fruitful way to deal with the first question raised in Subsec. 2.2.7.

---

[152] Plato, *Parmenides*, 132b-135c for a late statement of ther problem.

[153] It is therefore unfortunate when his philosophy is judged to be a castle in the air (Draper 1875, p. 26).

[154] Kepler 1596.

[155] Kepler 1609; 1618-1621.

3.3.4 *The Block Universe*

The kind of integration of random happenings envisaged by the third principle enables the fundamental *openness* of the dynamic processes of our world: any necessity is a consequence and not an antecedence of those processes (Subsec. 3.3.1). This view, which can be considered as a way to deal with the last philosophical question raised in Subsec. 2.2.7, is also in agreement with several acknowledged facts and theories, like evolutionary and epigenetic processes, quantum mechanics, the theory of complex and chaotic systems, and the observed time asymmetry of our universe. As to the latter point, as far as we know, it is a fact that the universe originated with an initial event called the Big Bang from which, in successive steps, the structure of matter, stars and galaxies, and finally planets able to host life have emerged. However, this third principle could be less in accordance with other scientific theories or results. If this were proved to be true, we would not be justified in formulating such a general heuristic principle of philosophy of nature. Therefore, this problem requires careful scrutiny. Let us then consider the theory in which a different understanding of our universe has been proposed. Relativity theory has often been interpreted as offering a fully deterministic view of our universe. In this subsection we only aim at critically examine this philosophical interpretation and not at discussing the relativity theory as such, neither we claim to summarize it in few words. As is well known[156], special relativity treats the speed of light as a constant and establishes specific requirements on the possible causal connections among events (Subsec. 3.2.4): in order for a certain message to reach the event E (which in turn is the minimal requirement for it to be able to causally influence E), the message must fall within, or – if it travels at the speed of light – on the surface (which is represented by light propagation) of, the so-called *past light cone* of E. The past light cone embraces all of the events in the past that could have dynamically influenced E. Reciprocally, any possible system of the universe that may be reached (or dynamically influenced) by any signal (or effect) that E sends (or determines) must lie within, or on the surface of, the *future light cone* of E. Moreover, general relativity is based on the principle of equivalence, according to which a uniform gravitational field is equivalent to a uniform acceleration. General relativity deals with the distortions that the presence of masses induces on time and space, generating different kinds of curvatures and affecting

---

[156] Penrose 2004, Ch. 17 for a summary.

the communication between different systems[157].

In both theories we have a single space-time and no longer a space separated from time or vice versa. Now, some scholars like H. Weyl have drawn the conclusion that a structure may be assigned to the universe such that it is objectively significant to say of any two different events that they are happening at the same time (at different places).[158] This seems to imply that the whole universe's space-time can be considered as a single atemporal block in which the asymmetric "flow" of time is only an illusion. Indeed, Weyl says that the inertial and causal structure, and therefore also the metrical structure, of space-time have a rigid character and are fixed once and for all[159]. Weyl grounds such a view on two assumptions[160]: (1) the state of a system at any moment uniquely determines its entire history, and (2) two cyclic events beginning in contiguous space-time can be arranged to end simultaneously. The first assumption is clearly violated by quantum mechanics (due to the occurring of random events) while the second cannot be generalized to the whole universe at any level of organization, since complex and chaotic systems show specific arrhythmias. However, there are also no a-priori reasons why current interpretations of quantum mechanics or complexity theory should be right and Weyl's conclusion wrong.

Thus, if Weyl is right, there are no random events at all but the whole universe is a single correlated network of causal relations in which any space-time point is directly or indirectly connected with any other (this would clearly represent a sort of overweight of our second principle although dynamically interpreted). Some doubts, however, can be cast on this worldview from the inside. First of all, relativity theory presupposes that we cannot have a general concept of simultaneity, such that events that are considered to be simultaneous in a certain reference frame may not be so in others. For this reason, when we say that something exists *now*, we are not allowed to transform such a temporalized mode of being into an atemporal one. In other words, to be temporally (or space-timely) determined does not imply being predetermined[161]. This is also evident by considering that proper time is another relativis-

---

[157] Thorne 1994.

[158] Weyl 1949, p. 95.

[159] Weyl 1949, p. 104.

[160] Weyl 1949, p. 103.

[161] As Torretti (1983, pp. 149-151) nicely shows criticizing Rietdijk and Putnam.

tic invariant[162]. A system traces a world-line in space-time, that is, a trajectory that describes its space-time evolution. Its proper time is the time measured by a clock that is carried by that system. In other words, different observers should agree about the time elapsed between two events along that world-line *on that* clock, even if they disagreed on the interval measured on *their own* clocks. Obviously, they would agree on the interval as far as it can be measured, and whose results can be recorded and therefore made accessible to further observers; this, however, has no significance for any single event (not happening in an interval). Therefore, any "here-now" is necessarily localised.

H. Bondi[163] raised the question that special relativity necessarily requires that there are world-lines that are not allowed since they fall outside of the light cones for a certain event E. From this he concluded that there must be random events in our universe that do not have causal connections with previous happenings, otherwise every event could be causally connected, at least indirectly, with any other event. In this way, light cones would not impose constraints on the causal interconnections of our universe but would simply describe the propagation of signals allowing the possibility that (in a finite or limited universe) over a long period of time everything can receive inputs from everywhere. Independently, G. Hegerfeldt proved that a strict localization of particles subjected to both quantum-mechanical and relativistic constraints contradicts the possibility of generalized causal connections[164]. As a matter of fact, J. J. C. Smart noticed that even in a Laplacian world (Subsec. 3.1.1) from a limited region of space we can deduce (retrodict) a great deal of the past history of the universe, while for performing similarly powerful predictions about future events we must consider initial conditions over a very wide region of space which for him justified the observed time asymmetry[165]. The reason seems to be quite simple: the events belonging to the absolute past of the event E could be (more or less) uniquely specified in records which are *part of that localized space-time region*, whereas events belonging to the absolute future of E are not, and therefore could well be unpredictable events in the sense of quantum mechanics[166] or even impossible events in the sense clarified by Bondi. This means that it is again the kinds of dynamical effects that a certain event induces on its sur-

---

[162] Zweibach 2004, p. 25.
[163] Bondi 1952.
[164] Hegerfeldt 1974; 1985.
[165] Smart 1954.
[166] Grünbaum 1973, pp. 322-324.

roundings that determine time asymmetry and the related openness of dynamical processes. The primacy of dynamics on a static world-picture can also be seen from another point of view. Since no signal can travel faster than light, the concept of rigid bodies breaks down and we cannot assume that the world is constituted of bricks[167] (not even of space-time bricks).

It is worth mentioning that one of the maximal experts of general relativity, J. A. Wheeler, in about 1972 shifted from a position considering space-time as the basic ontological stuff, to viewing it as an abstraction from quantum events (see also Subsec. 3.2.4 on the possible emergence of space-time from quantum information) thus proposing a relational worldview[168]. Wheeler was well aware that quantum correlations are fully independent of space-time, so that neither predictions nor retrodictions are possible for quantum events. Now, quantum events establish an immediate, point-like determination that eventually allows for *certain* predictions or retrodictions[169]. Some of these events then allow for establishing certain space-time (and causal) interdependencies. However, the latter cannot be indiscriminately generalized to the whole universe, so that the space-time metric should be rather considered as local or at least strongly limited by quantum-mechanical events.

It is true that in gravitation (general relativity), the so-called gauge invariance is translated in a parameterisation invariance, which more simply means that the choice of the coordinate system used to parameterise space-time does not affect the physics[170]. Should this invariance also imply that space-time metric is universal or global? A reasonable answer could be that any parameterisation is good for any sufficiently large region of causally interconnected events. As a matter of fact, there is no univocal (Lorentz-invariant) space-time connection between space-like events, namely events that cannot be connected by a light signal (they fall outside of the light cones of each other), and therefore different observers cannot agree on any interconnection between them. The Lorentz-invariance involved in the relation between two space-like events is taken to be the proper length, that is, the measure of the relativistic distance between them. However, in order to deal with this length it is necessary that a signal starting from each of the two events reaches an observer. If so, these two events have been integrated into the

---

[167] Bohm 1980, p. 9.
[168] See Grünbaum 1973, p. 729 for a short account.
[169] Wheeler 1983.
[170] Zwiebach 2004, pp. 53 and 181.

network of interconnections of this and similar observers (and they have also lost their irreducible nature of here-now). Therefore, if we would like to avoid a possible conflict between general and special relativity, it is suitable to interpret the parameterisation invariance as saying that every time we measure, or interact with, other systems in our universe, these causal interconnections are independent of the choice of coordinates. However in order to do so, we can no longer treat events as irreducible happenings but should insert their effects into a web of interconnections constituting a necessarily regional "piece of the world"[171].

For these reasons, Wheeler spoke of a structure of space-time made of different "leaves", whose tops can represent events of registration while lower ends can be the beginning of some dynamical open process. Wheeler called such a world "participatory": every property manifests itself through an event that is the result of an interaction process of a system with something else. We know very well that the second major worry of Einstein after determinism was to ensure the independence of the physical reality from any observer[172], as it is expressed by his criterion of reality (Subsecs. 2.2.6 and 2.3.5). We should avoid falling into any form of subjectivism or idealism. However, Wheeler's proposal allows for finding an interesting compromise: reality is interdependent without depending on any observer. On the contrary, the web of the world's relations and interdependencies is somehow objective and real, although not absolute or universal in any sense due to the occurring of local happenings or physical processes (therefore, this can be considered to a certain extent as dealing with the issue of the ontological nature of relations, raised as the 5th question in Subsec. 2.2.7). This may even help us to distinguish between a formal and a physical world. Relations could be taken to be absolute only as objects of mathematical or logical statements, that is, without taking into account those local physical processes that confer also to them a kind of objectivity.

There is an important historical connection with Leibniz[173] who stressed very much the idea of a perspective-like universe. For this reason, Leibniz criticized Newton's theory of an absolute space and time (Subsec. 2.3.3) and can be considered as a precursor of relativity theory: as a matter of fact, when Mach subjected

---

[171] See Auletta 2011a, Subsec. 2.3.1 and Sec. 25.1, where it is shown that this and similar circumstances depend on an information accessibility principle.
[172] Jammer 1999, p. 71.
[173] Leibniz 1686b, p. 434; 1702, p. 564; 1712-1714, § 57.

Newton's assumptions to a sharp criticism, he took Leibniz's standpoint into account[174]. The problem with Leibniz seems to be that he only acknowledged correlations (but not causal relations) and also assumed that everything is connected with everything else, a sort of non-dynamical block universe (pre-established harmony). This is quite understandable if we acknowledge Leibniz to be the father of the second principle above (Subsec. 3.2.3).

3.3.5 *Emergent Monism*

Our third principle shows that nature is essentially a source of novelties while leading to the instantiation of new stable solutions thus promoting the emergence of new forms of complexity at any level[175]. This whole universal process of arising variety may be called *emergent monism*[176], in accordance with our understanding of naturalism (Subsec. 2.2.5). We stress here that the term "monism" should not be understood as a variant of materialism. We have indeed pointed out that already at the physical level we should consider formal constraints and not only traditional physical parameters. When we speak of monism, we rather wish to focus on the *recurrence* of the same fundamental behaviours (as expressed by our three heuristic principles) across all domains of nature. In particular, the regularity expressed in the second principle tells us that correlations are established at any level of complexity. On the other hand, the term "emergent" stresses the extraordinary variability of nature, as expressed by the first principle, and the impossibility of reducing what is more complex to what is less complex[177]. As we have already mentioned in the introduction to this section, to maintain these two aspects alone would eventually lead to some form of dualism. In order to elaborate the idea of an emergent monism we need a true conceptual mediation between these two aspects. This is precisely provided by the third principle, which allows us to establish a dynamical continuity in the variety across all the levels of complexity[178].

---

[174] Mach 1883.

[175] Clayton–Davies 2006.

[176] Auletta 2011a, Sec. 2.4.2.

[177] Coyne 2005.

[178] Leibniz stressed several times that the beauty of our world stems from the combination of the simplicity of natural laws (often called hypotheses) and the richness of phenomena (for instance, 1686b, pp. 431).

We mention here that this difficult issue can be enlightened when we consider its relations with information. We have indeed remarked that correlations and formal constraints are mathematically treatable in the language of information (Subsec. 3.2.4). We have also remarked that organisms aiming at controlling environmental parameters perceive environmental fluctuations in terms of informational surprisal[179] (Subsec. 3.3.3). Surprisal is actually a general concept that enters not only into biology but also into any exchange of information, even at the physical level. In other words, any unpredictable event represents a surprisal relative to the equilibrium state of a system or to what is expected. When certain systems, especially when complex, emerge, they undergo a process in which some informational connections between components are reinforced, whilst others are silenced, and this often happens according to chance events that are produced along the process. This is what occurs during the maturation of the brain (Subsec. 3.1.6) and more generally during epigenetic processes (Subsec. 3.1.5, 3.2.5 and 3.3.1). These processes can be understood as a dynamics through which certain information is selected and acquired in response to certain external stimuli and by making use of information shared between the sub-components of the system. If information is really a basic category of our universe, this approach can turn out to be very fruitful in trying to understand emergent monism[180]. This has connections with both the 5th question (especially in relation with the notion of system and its intrinsic dynamical aspects) and the 7th one raised in Subsec. 2.2.7.

Finally, we recall that the idea of emergent monism has been originally introduced by Lloyd Morgan[181], and interestingly reprised by P. Clayton in a theological effort dedicated to uphold the theoretical emergentist view of the natural world on the basis of contemporary science, and ultimately to assess this view in relation to transcendence[182]. In the case of the emergence of the mind, Clayton remarks that emergent monism avoids the traditional substance dualism by considering mental properties as depending on the lower level of the neural hierarchy, yet genuinely emergent from it: "Thought is dependent on neurophysiology but not reducible to it". We recall here that this view is fully in accordance with the traditional Aristotelian standpoint. According to the modern version of this standpoint, conscious experience in its peculiarity is one of the intriguing new properties emerged in the

---

[179] Auletta 2011b.
[180] For readers interested to deepen this dimension, see Auletta 2011a.
[181] Lloyd Morgan 1923.
[182] Clayton 2004.

course of natural history within the one and the same natural world. It is very important to clarify that although we attribute a certain activity to the mind, the latter cannot be understood as something that dynamically causes brain excitation patterns from above. On the contrary, the mind rather has a constraining influence that, when combined with events occurring in neural excitation patterns as well as in the external world, enables new forms of complex causation in which a dynamism is established thanks to the concourse of the mentioned physical factors. This also helps us to avoid scientific mistakes and philosophical misconceptions such as those that invoke a dynamic-causal role of the mind on quantum systems, as mentioned in Subsec. 2.1.6.

3.3.6 *Evolution and Directionality*

Let us now come back to a specific application of the previous ideas and then try a further generalization. Biological evolution seems to have led to forms of life that are more and more able to exert an information-control on their environment, i.e. to monitor certain parameters, to change some environmental factors, and to "subdue" (up to a certain point) their immediate surroundings to their vital needs[183]. As we have seen before, organisms are able to activate new and canalizing constraints that have effects on themselves and on the environment (Subsecs. 3.3.1 and 3.3.5). The particular evolutionary paths that were actually followed may have been extremely different, since contingent irreducible happenings, encompassing both genetic and phenotypic variability as well as environmental fluctuations, copiously occur. So, the *specific and detailed* solutions that are produced along the evolutionary paths are obviously not previously fixed or necessary, but the convergence among many of those solutions unveils the fundamental fact that further evolutionary steps *need* to find new levels of order and stability, which are rooted in the principles of entropy and information that were stated above (Subsec. 3.2.4).

The term "need" that we have just used could be considered to be inappropriate. As a matter of fact, nature shows an overall tendency to disorder expressed in entropy growing (this is the content of the second principle of thermodynamics). Such a tendency grounds the irreversibility of the occurring dynamic processes. Therefore, the fact that we also observe continuous manifestations of order at all levels of our universe provokes astonishment. How is it possible? Also in quantum

---

[183] Auletta *et al.* 2008; Auletta 2011a, Chs 8-11.

mechanics irreversible processes occur, whose paradigm is represented by measurement. However, a fundamental theorem of quantum mechanics tells us that for any irreversible process we may find a larger context to embed it that is reversible as a whole[184]. Therefore, if the entire universe obeys quantum-mechanical laws (as we are inclined to think), we would expect that the tendency to disorder is continuously balanced by a tendency to order, so that the net result is precisely zero. As a matter of fact, an isolated quantum system shows zero entropy, which goes together with reversibility. The conservation of both order and disorder in our universe is likely to be a far more general principle than the conservation of physical quantities like mass, energy, and momentum. There are situations in which the conservation of energy can be "violated" like in the "creation" of virtual particles in a vacuum field. Moreover, quantum mechanics does not obey the principle of least action that is connected with those conservation laws. Obviously, the tendency to disorder is spontaneous and in this sense more fundamental. Indeed, Leibniz and the fathers of classical mechanics considered that nature always "chooses" the easiest solutions, and disorder is easier than order. This principle has a counterpart in the current developments of statistical mechanics according to which the possible disordered states of a system are much more than the ordered ones (for instance, there are quite likely to be an infinite number of ways to break a cup, but only a few to build it). The crucial point is that every time such a tendency manifests itself (and assuming that we are right in the previous assumptions) a *compensatory* tendency to order is also likely to be produced to preserve this net balance. This second tendency can be said to be less fundamental and not spontaneous, but forced by the first one. Thus, we see again that a certain necessity likely arises in our universe only as a consequence and does not determine further developments as an antecedence. This may also clarify why any equilibrium reached by all the systems in our universe is somehow unstable (Subsec. 3.3.3), and why the last word is dynamism. A consequence of these considerations is that the tendency to order will also be displayed through growing levels of complexification, without resorting to any form of vitalism or extra-natural factors. In order to understand this, it is sufficient to take into account the elementary consideration that emergent realities show a higher number of constraints than lower levels of reality do. In other words, we cannot say, in any meaningful sense, that evolution has been targeted in advance[185]; however, we can very well understand how the emergence of some more advanced forms of life, (i.e.

---

[184] Auletta *et al.* 2009a, Secs. 17.7-17.8.
[185] Stoeger 1998.

those that are able to deal with the environment in richer forms than their predecessors), may be considered as a consequence of a combination of selective pressures and constraints. Therefore, biological evolution, like cosmic evolution, can show a certain directionality without any finality[186]. Indeed, this directionality is only a natural *consequence* of the basic principles that we have considered above, and not something primary or designed in advance.

The sketched explanation of how both variety and order are built and even increase across the evolutionary history of our planet clearly shows the inconsistency of a theology of nature like that of Paley or of the Intelligent Design supporters, who aim at inferring a purpose and design of the biological world and evolution from the ordered complexity of living beings (Subsec. 2.3.3). It is also true that we have distinguished between two different approaches of natural theology, one (also followed by the current supporters of Intelligent Design) focusses on design as an explanation of *particular* aspects or occurrences in our world, and the other is centred on the *overall* features and laws of our universe. If we take natural theology in the latter sense, it is obvious that these laws and characters have only got a general value, and we think that most authors in this field use the word "design" in this sense. In fact, what is really amazing about our universe is the incredible beauty, simplicity and order of many natural phenomena and even of its overall constitution[187]. This character of the world has always inspired not only artists but also scientists, as we have seen in the case of Einstein (Subsecs. 2.3.5 and 3.2.6). Obviously, the awareness of the beauty and order of the universe is also fundamental for religious belief and for theological investigation (as we shall see in the next chapter). This may well explain why people supporting natural theology have drawn the pivotal lesson of a divine action moulding brute matter and conferring to it those general properties and laws that we admire (a vision ultimately stemming from Plato's Demiurge). However, if another general character of our universe is the continuous emergence of novelties (according to our first heuristic principle), the word "design" seems inappropriate. We have indeed stressed that emergent phenomena are the result of convergence processes. As already mentioned, we think that the harmony and regularity that is shown at all scales of our universe need not have been designed in advance, but is actually well

---

[186] Auletta-Stoeger 2010. Edwards 2010, pp. 8-11.
[187] Holder 2004

explainable according to our third heuristic principle, as the result of a dynamism in which many different factors need to be put in accordance in order to arrive at a relative stable solution. This means that what allows for these convergences is always a selective fork: all the solutions that are unstable or do not fit previously established conditions and constraints are discarded. As we shall show in the following, this way of understanding how order and beauty are produced in our universe is much more theologically interesting and even consonant with the idea that we have of God, rather than models of divine action that are ultimately based upon an analogy with human watchmakers[188].

It is appropriate to consider here the more radical question (Why is there something rather than nothing?) listed in Subsec. 2.3.7: a typical problem at the crossroad of metaphysics and theology of Creation. Indeed, having discarded the design option, we can consider whether or not the universe as a whole is the result of chance only. In Subsec. 2.2.7, we have recalled Deutsch's examination about the possibility of a first principle or law explaining everything. Recently, another physicist, V. Vedral, had raised the same question by asking whether or not there is a law without law, that is, a self-founding law[189]. He founds this principle by assuming that information is the ground of everything. We have also seen that information is crucial in our universe (Subsecs. 3.2.4 and 3.3.5). The author says that assuming a universe with an initial state of zero entropy at time 0 basically implies that everything was possible at this stage but nothing was realized: a sort of quantum "symmetry" (superposition) of all possible states of the universe. According to Vedral, this would coincide with the state of absolute nothingness. In this way, the problem of the so-called initial conditions is solved (we do not need to assume any particular initial state), a point that also S. Hawking stresses in his recent book[190]. The beginning of the universe would have occurred out of this initial nothingness through a random fluctuation that does not need causes as such. As we have seen, there are good reasons to think that quantum systems do indeed make such random jumps. The subsequent evolution of the universe would have consisted in further "symmetry breaks" or further random happenings that are able to determine the universe more and more as the sculptor makes a sculp-

---

[188] Which was also Newman's opinion as it is reported in Hughes 2009, p. 209. See Stoeger 1998.

[189] Vedral 2010, Ch. 12.

[190] Hawking-Mlodinow 2010, pp. 133-44.

ture in a process of progressive plastic fine-tuning (it is the author himself who uses this metaphor of a moulding activity). Such a process would be in accordance with the second law of thermodynamics because it implies an increase in disorder (entropy), manifested by an irreversible sequence of "choices". In this way, reality would have spontaneously emerged from nothingness. This account is clearly fascinating. However, there are some drawbacks. The author himself acknowledges what we have just mentioned, i.e. that according to quantum mechanics for any entropy-increasing process a larger system can be found in which entropy is preserved in a reversible way so that its increase is only a local process. Therefore, if the whole universe has started with a zero-entropy state (which is an initial condition[191]), according to quantum mechanics it would preserve such a state forever (which would be in accordance with an adiabatic expansion of the universe[192]). Moreover, to assume both (i) that we had a state of the universe at the beginning in which any configuration was possible and (ii) that we get some determined universe as a reduction-process from this plenitude of possibilities means to assume that the universe is ruled by the same quantum-mechanical principles that we apply to local measurements. Is this a correct step? And where did these principles come from? Furthermore, to assume that nothingness is equal to a zero-entropy state is a huge step, since zero entropy means maximal order. Why should this state be self-justified? Why is there no initial state with less order or even with maximal disorder? If nature follows the easiest solutions (the author assumes an entropy-growing process for this reason), should not the latter be easier? The above assumptions could be justified by already implicitly assuming the validity of quantum mechanics, since we expect that an isolated quantum system has indeed got zero entropy. Moreover, an initial superposition of every possible state is not nothing but is precisely the definition of an isolated quantum system. Where does this quantum system come from? As a matter of fact, the physicist who originally introduced the hypothesis of an initial quantum fluctuation, namely E. Tryon[193], considered that a

---

[191] Indeed, he seems not to ask what the intial entropy of the universe was or what the entropy of the vacuum was. If he had raised such a question, he would have seen that the problem of the initial conditions is inescapable.

[192] Mo *et al.* 2010, pp. 108-110 and 129-133. See also Penrose 2010, although we do not agree with his general hypothesis.

[193] Tryon 1973. It was this paper that introduced the notion of quantum fluctuation as a *creation* from nothing.

necessary requirement of this model was to assume that the universe was, and is, embedded in an enlarged (vacuum) space. Such as space is not nothing. Where did it come from? These are some of the problems that we see in such a proposal and which can be dealt with thanks to a correct philosophical and theological approach.

3.3.7 *The Third Principle on a Philosophical Background*

Having overcome some fundamental possible objections derived from natural sciences and dealt with some theological troubles, we turn to some philosophical aspects of the third principle. We are not far away from the truth when we say that dynamicity is a cornerstone of Aristotelian philosophy. The main goal of Aristotle's philosophy of nature is to understand how change and dynamism is produced in finite beings. Unfortunately, his theory was strongly misunderstood at the beginning of the Modern Ages when the formal causes were mixed with efficient causes[194] (Subsec. 3.2.5). This is quite natural because, as we have stressed, modern science began with mechanics. Now, in a mechanical framework, categories such as *potentiality* or *actualization* lose any sense. The problem here is that one often forgets to consider that Aristotle's philosophy of nature was not addressed at mechanical systems at all, but essentially targeted biological ones[195]. As a matter of fact, many of his ideas turn out to be particularly interesting in biology today, especially when epigenetic processes are taken into account. It is not by chance that we considered epigeny as one of the main application fields of the third principle (Subsec. 3.3.1). We recall that in the book on animal parts, Aristotle considers epigeny as a dynamical process in which a species-specific form is built in a stepwise fashion by starting from an embryonic stage[196]. In order to understand the issue here at stake, we may compare this vision with the mechanist view: the latter necessarily implies a biological preformationism (Subsec. 2.3.3), i.e. the idea that a sort of *homunculus*, of which the mature phenotype is only an expanded copy, is

---

[194] Pasnau 2004. However, the author also shows how complex the position of a scientist like Robert Boyle is, since he deals with the problem of the conformation or texture of certain bodies, an issue that has never been completely solved in modern science.

[195] Unfortunately, this is a misunderstanding that still dominates the way in which many scientists consider Aristotle. For instance, in Hawking-Mlodinow 2010, p. 24, he is still considered as the proponent of a physical theory that is alternative to the one of the Modern-Ages. This is also related to the fact that the authors show little awareness of biological problems.

[196] Aristotle *Part. An.*, Book II, Ch. 2.

contained in the egg or the sperm. It is difficult for us today to grasp how much this vision of biology has hindered true progress in the theory of development for centuries. Apart from a few exceptions (like K. von Baer), most biologists have followed this line of enquiry, and we may even consider the view, supported for more than half a century, that the genetic system has a blueprint of the organism as an updated version of preformationism. In other words, we are inclined to accept that Aristotle's *Physics* turned out to be not very helpful for the rising of classical mechanics, but we also emphasize that the mechanist view of the world induced by the latter has hindered a proper development of biology.

Given these premises, it is important to have a look at Aristotle's theory of dynamics in its generality. Aristotle considered a large variety of changes, certainly wider than it is admitted in mechanics but again very interesting when looking at epigenetic processes. Sometimes he says that change encompasses change of form, growth and diminution, combination of pre-existing components, and modifications of the involved material[197]. We do not need to insist here on the four canonical forms of change: of substance, quality, quantity, and location[198]. As it is well known, Aristotle understood dynamics as involving potential elements and resulting in a final state of actuality[199]. He uses three fundamental terms[200]: *dynamis* for understanding potentiality, *entelecheia* meaning the final state that ends the dynamical process, and *energeia* as the dynamical trade-off between the two[201]. We have also stressed that our principle of dynamicity is focussed on a dynamical interplay involving both potential elements like correlations and constraints (Subsec. 3.2.5), and actual events (Subsecs. 3.1.5-3.1.6).

In this context, we also emphasize that the formal constraints we have introduced in connection to the second principle (Subsec. 3.2.4) are linked to both Aristotle's formal causes[202] and his idea of form (we have indeed stressed that information is essentially a formal quantity).[203] However, it is also true that Aristotle established

---

[197] Aristotle *Phys.*, 190b 5-10.

[198] Aristotle *Phys.*, 200b33-201a19.

[199] Aristotle *Phys.*, 201a10-19.

[200] Aristotle *Phys.*, 190b30, 191b35-192a14, and 198a30-35. Blair 1992. Aristotle's reflections on time as dynamical (Aristotle *Phys.*, 222a10-20) is also worth considering.

[201] Aristotle *Phys.*, 190b30-32.

[202] Aristotle *Phys.* 194b24-195a3 and 198b1-9; *Met.* 983a24-34.

[203] We should not forget that Aristotle uses the term *morphe* in order to denote natural forms instead of *eidos* as Plato did.

a tight connection between formal and final causes since he understood the former in the sense of the form to which a given substance tends[204]. Obviously, this may be true when, during the epigenetic process, a species-specific form of the organism is realized, but cannot be generalized to the whole of nature nor to the whole of biology[205]. It is possible that this is the result of a misleading reading of Aristotle's conception of the relation between final and formal causes, since Aristotle himself thought about the so-called substantial form rather in functional terms[206], an approach that we fully support, especially because it is potentially very fruitful from both the scientific and the philosophical points of view. It may be that Aristotle's original aim was to show that the realization of a form displays an intrinsic teleology only in the case of the species-specific development of an organism[207]. Then, the notion of constraint we adopt here is more general than that of form understood in the strict Aristotelian sense. It is finally worth emphasizing that the Aristotelian concept of material cause may be somehow associated with our first principle since, to our understanding, the variety of material structures is a necessary condition for the emergence of higher levels of complexity (Subsec. 3.3.2).

A second major philosopher who dealt profoundly with the issue raised in this section is Charles Pierce. Actually, our dynamicity principle is strongly indebted to his category of *thirdness*. He stressed that this category should indeed be understood as a possible half-way between the first two aspects of nature, bridging the chasm between variety and interdependence, and also as a general tendency of nature to acquire habits[208]. Moreover, for Peirce a kind of principle of dynamicity was the cornerstone of pragmatism. He affirmed that our concept of an object is the sum of all possible effects that this object can determine on the world[209], which

---

[204] Aristotle *Phys.*, 198a25-29.

[205] Aristotle (*Phys.*, 199b15-20) says that natural things are those which move in virtue of a principle inherent in themselves towards a determined goal.

[206] As again shown by Pasnau 2004.

[207] Frede–Patzig 2001, p. 260.

[208] Peirce 1887-1888, pp. 172 and 208. Similar formulations can be found in *EP*, II, pp. 183 and 269 or *CP*, 1.328 and 1.337. Peirce is not always consistent with his formulations and we can also find different ones like in *CP*, 6.211.

[209] Peirce 1878b, p. 266; 1903a, p. 135; 1905a, p. 332; 1905b, pp. 346 and 358. This is in perfect agreement with Hamilton's (1859, III, Lect. VIII, pp. 135-136) view that a concept only represents a pontential but not actual universality as far as it *can* be applied to a certain number of objects (its extension). However, when it is actually applied, it only shows a special

is another way of saying that it is the sum of all interactions that it shall have with the world. It is worth mentioning here that Peirce also took into account an evolutionary dynamism in the universe that he called *agapism*[210].

We finally recall Whitehead's principle of process, according to which (finite) being can be identified with becoming[211]. Also some contemporary theologians have dealt with this issue. Maldamé stresses that nature is to be considered as essentially dynamical[212]. Tanzella-Nitti, recalling Prygogine, notes that self-organizing processes that are far from equilibrium display a delicate interplay between random fluctuations (our first principle) and deterministic laws (a decisive component of our second principle), and considers this as being a rather diffused character of natural reality[213]. We shall come back to further possible theological implications in the next chapter.

3.3.8 *Brief Epistemological Consideration*

Approaching the problem from the *epistemological* side, it may be shown that the current efforts of natural sciences should not be directed towards an unlikely-to-be-reached prediction of any *particular* happening or of the single pathway followed by natural phenomena; rather, according to our no-go theorem (Subsec. 3.3.3), they should be addressed at the general constraints which allow one or another happening or pathway to occur, as well as at the final result that all of these different occurrences bring about[214]. One might take this as a limitation of the scope of science with respect to what was assumed in the 19th century, when classical physics formulated the all-embracing rational mechanics, according to which any natural phenomenon would be fully predictable (Subsec. 3.1.1). However, the possibility of conceiving such a complete knowledge stemmed from the fact that the considered domain of reality was narrower than the one currently investigated, and it was often even studied in an idealized way.

---

form of attribution and not a universal one. For instance, one cannot actually conceive a triangle without representing it, e.g. as rectangle or equilateral. This point was also well understood by Berkeley (1709, §§ 123-127; see also Auletta 2007c).

[210] Peirce 1893.

[211] Whitehead 1929, p. 213.

[212] Maldamé 2003, pp. 248-251.

[213] Tanzella-Nitti 1997, p. 24.

[214] Parisi 1999.

Let us stress that the general character of current scientific explanations does not imply that particular happenings should not be taken into account either. Indeed, when we address the domain of life, or that of cognition, specific *occurred* happenings in the history of our universe have to be considered as the perturbations that have activated formal constraints, and led, in a dynamic integration process, to the very emergence of those new levels of complexity. This emergent complexity is displayed through properties and behaviors that cannot be reduced to properties and behaviors of the systems out of which this complexity came out. Such a property or behavior is then robust to fluctuations of variables or conditions out of which is emerged, becoming in this way independent of the details determining the latter[215]. Therefore, the principle of the dynamic interplay could help scientific enquiry to mould riper explanations that are able to deal with a universe that we discover to be growingly complex. This qualifies our previous statement that knowledge of more complex levels of reality could lead to a better understanding of more basic ones (Subsec. 2.2.4).

3.4 Philosophical Reflection on the Three Principles

As a conclusion of this extensive enquiry and after the examination of several philosophical standpoints that are somehow connected with our formulation of the three philosophical principles, let us focus on some general aspects of the problem. As we have already mentioned (Subsec. 2.2.6), these principles have a heuristic character. This means that we postulate them in order to stimulate and promote future research. So, we are interested in testing how fertile such principles can be without transforming them in a closed conceptual system for classifying already acknowledged results or facts. As far as we can understand, the attitude to build a closed system portrays Hegel's use of his three fundamental ontological categories: being in itself (thesis), being other (antithesis), and being in and for itself (synthesis).[216] These categories bear a certain resemblance with our three principles, but with many important differences. Indeed, Hegel understood the first category not as meaning a true random generation of variety, but in the sense that what appears to be immediate and therefore not rationally (dialectically) justifiable will be shown to be actually deduced in the further dialectic process[217]. Moreover,

---

[215] Batterman 2002.
[216] Hegel 1807, p. 28; 1833, v. 2, pp. 483-506.
[217] Hegel 1833, v. 1, pp. 55-58.

the very beginning of science, according to Hegel, namely the absolute immediacy as such, is not a particular event, or even the general notion of event, but something that has no determination, no content and therefore is absolutely abstract, i.e. pure Being[218]. To a certain extent, we may consider this position as the opposite of our first principle, since it denies the irreducibility of the variety that we meet in our experience. Such an experience cannot be mixed up with an intellectual reflection (which is involved in Hegel's deduction), and we have indeed stressed (introduction to Sec. 3.2) that primary experience of events are not rationally articulated. This also implies that while our first principle, as inspired by C. Peirce's firstness[219] (Subsec. 3.1.8), does not explain any single happening itself but only the general fact of the occurring of irregularity in our world, Hegel sometimes aims at transforming brute facts into rational laws. It may suffice here to think about his rational justification of the Prussian state (including the monarchic form, the censure, the corporations and castes) in the *Philosophy of Right*[220]. Moreover, Hegel understood his second category essentially in terms of the suppression of the apparent immediacy of the first moment (Subsec. 3.1.1), which implies a fundamental negative and dynamical component of the second dialectical step[221], which is not very much in accordance with our second principle (Subsec. 3.2.3). Finally, Hegel's third category claims to be a final synthesis through which the previous conflicts are completely resolved and the previous categories fully abrogated in their autonomy[222], whilst we have stressed the intrinsic openness of the principle of dynamicity (Subsec. 3.3.3). Indeed, for Hegel what is true is only the whole. This is due to the fact that his philosophy portrays the self-movement of the Absolute or its immanent activity, and therefore it is a *necessary* development[223]. Finally, we have anticipated that Hegel did not use his categories in a heuristic way but, quite the opposite, as containers in which he claimed to frame any possible aspect not only of our universe but even of being and of rationality. Therefore, the issue at stake here is: can general categories be found in a process of empirical in-

---

[218] Hegel 1833, v. 1, pp. 54 and 66-67.

[219] Peirce 1892a, p. 308.

[220] Hegel 1821, §§ 298-320.

[221] Hegel 1807, p. 36; 1833, v. 1, pp. 6 and 26-27. It is likely that these formulations of Hegel may have influenced C. Peirce since we have remarked a certain incoherence between some of his statements, and in particular we have emphasized that he sometimes interprets it in dynamical and negative terms and not as being focussed on the universe's interdependencies.

[222] Hegel 1833, v. 1, pp. 93-95 and 139.

[223] Hegel 1833, v. 1, pp. 7 and 9; 1807, p. 24.

vestigation (or at least in connection with the latter), or do they need to be assumed a-priori and the contents only afterwards deduced from, or at least framed in, them?

This is one of the most difficult philosophical problems. In order to advance consistently in this enquiry, it is very important to settle our own standpoint as clearly as possible. We consider ourselves to be followers of Aristotle's methodology, according to which philosophy consists in the search for causes and explanations (Subsec. 2.1.1), as he clearly states in his *Methaphysics*[224]. Aristotle emphasizes that the path of investigation must go from what is more immediately cognizable and clear to us to what is clearer and intimately cognizable in its own nature, a standpoint that has also been assumed by St Thomas (Sec. 1.2).[225] This is why *Metaphysics* comes conceptually after *Physics*[226]. Moreover, Aristotle states that the things that are most obvious and immediately cognizable by us are concrete and particular rather than abstract and general, so that elements and principles are only accessible to us afterwards[227], in accordance with our own standpoint expressed in Subsecs. 2.1.2-2.1.4. We do not need to recall here the empirical studies of Aristotle in biology, especially concerning animal anatomy (Subsec. 3.3.7). This standpoint may well be considered as a mediation between empirical elements, representing the immediate contents of cognition, and reflected knowledge that is the specific contribution of what we call philosophy today. As a matter of fact, Aristotle's logic can be understood as an attempt at finding a middle term between general hypotheses and experience[228]. We think that it is on this line that St Thomas said that Aristotle's standpoint should correct Plato's one: although the former agreed with the latter in distinguishing intellection from sensation, Aristotle was nevertheless able to emphasize that any operation of the mind must be triggered by sensation and physical interaction with the material world[229].

Now, the question we would like to raise is the following: why in modern times did continental European philosophy not follow this approach and finally detach itself from scientific developments, consequently not taking into account the empirical contents deriving from scientific investigations (contrarily to the empiricist Anglo-Saxon

---

[224] Aristotle *Met.*, 994 a 19-30 and 995 a 15-20.

[225] In this context, we may also recall the enlightnening distinction between assent to concrete objects and assent to abstract notions drawn by J. H. Newman (1870, Ch. 4). Newman made clear that we may still confuse assent to notions with inference (Subsec. 2.2.3), but this confusion never occurs when we deal with the objects given in our experience.

[226] Averroes, *AM*, p. 26 (j34).

[227] Aristotle *Phys.*, 184 a 10-b 14.

[228] Auletta 2011d and literature therein

[229] Aquinas, *S. Th.*, I, q. 84, a. 6.

tradition as expressed by philosophers likes J. Locke and D. Hume)? This is especially astonishing when considering that many continental philosophers have strongly contributed to the progress of science. The best answer that we can provide to the previous question is that continental modern philosophy started with, or at least had an important climax in, Descartes' thought[230]. Now, the whole analysis in his metaphysical meditations is characterized by introspection through which he tries to overcome some radical doubts and reach grounding certainty. This is an incredible turn in the history of thought since it gave rise to a fully new methodology in which science claims to be grounded on subjectivity, suspending (the so-called *epoché*) any possible empirical experience with an act of self-foundation that acknowledges the Ego as the only ultimate evidence, from which all other notions should be derived[231]. As pointed out by Hegel, modern Idealism, of which Descartes can be considered as the father or at least as a precursor, shows an important difference relative to the ancient Platonic Idealism: the latter was indeed a sort of objective Idealism (that is, of forms or ideas), while the modern variety is an Idealism of the subjectivity and of its rational activity[232]. This may also explain, at least in part, Leibniz's rationalist standpoint: he started his intellectual journey with the idea of a combinatorial art understood as an a-priori science able to derive any possible concept through combinatorial means[233]. This formal-logical approach accompanied him for his whole life as attested by his numerous projects for academies based on the idea of a universal science and a universal language of science (Subsec. 2.1.2). This may clarify why Leibniz, although being a great mathematician and physicist himself, was unable to establish a tighter connection between some of his brilliant philosophical insights and the scientific developments of his time, although both he and Descartes had a later and posthumous influx on the arising of Lagrangian mechanics in France at the end of the 18th century[234]. Moreover, such a

---

[230] Descartes 1641.

[231] Husserl 1950. We may remark that when Husserl (1950, § 6) looks for an absolute first evidence that cannot ever be doubted, he incurs in a misunderstanding that is typical of the foundationalist approach. This approach tries to find an ultimate and absolutely unquestionable positive evidence, while, according to us, evidence is by definition negative and therefore can only come from experience representing a counterexample to our theories and hypotheses (Subsecs. 2.1.3 and 3.1.2, Introduction to Sec. 3.2).

[232] Hegel 1833, v. 1, pp. 31-32 and 145-146.

[233] Leibniz 1666.

[234] Maiocchi 1995, pp. 352-353. Apart from historical reasons, we may also add that most parts of Leibniz's philosophy were expressed in an obscure language (for instance, his *Monadology*) that was barely understandable outside of very small groups.

philosophical trend also led to some typical scientific errors[235]. Peirce remarked that if we take fifty metaphysicians, each holding opinions that none of the other fortynine can admit, everybody will nevertheless generally regard his own opinion as being more certain than that the sun will rise tomorrow[236]; this disregard for others' opinions is an unfortunate consequence of the introspective withdrawal into the Ego. Even a philosopher who was so intensely focussed on the results of the natural sciences of his time, like I. Kant, ended up developing an a-priori deduction of the basic categories of human knowledge.

We are obviously not denying the richness of particular contributions and ideas of all the mentioned philosophers and of many others. A great thinker cannot be reduced or confined to schematic positions. We are rather emphasizing some general and common methodological characters that can be helpful when we wish to understand why European continental philosophy cut itself off from scientific developments. Moreover, as we shall see in the next two chapters, we uphold the fundamental insight of modern Idealism about the mind as activity. However, what we shall try to do is to separate this fundamental insight from foundationalist claims, especially when connected with a subjectivist standpoint. There is an important and instructive difference between the developments of science and philosophy that needs to be considered here. The root of the subjectivist approach in philosophy may be found in Reformed theology, which also influenced the very first stages of modern science. Indeed, P. Harrison has shown that the terms *experiment* or *experimental* may originally have meant *personal experience*, and only in subsequent developments (in a riper stage of science) have been interpreted in the same sense in which we understand them today[237]. It is likely that it was Francis Bacon[238] who provided the background for this turn (especially considering his influence on the incorporation of the Royal Society): instead of appealing to personal introspection in order to overcome the intrinsic limitations of the mind determined by original sin, he proposed a regimen based on some methodological constraints of an "external" kind relying finally on ex-

---

[235] Apart from the theory of vortices, we may indeed consider that Descartes tried to deduce a solution for the fall of bodies on the Earth from his philosophical speculations, in which there was no place for time due to his full geometrisation of physical reality in which only extension played a role (Koyré 1966, pp. 107-136).

[236] Peirce *CP* 1.32.

[237] Harrison 2007, pp. 125-138 and 141-154.

[238] Bacon 1620. Under this point of view the Baconian methodology is a form of error correction of the kind envisaged in Subsec. 2.1.3. See also Hamilton 1859, IV, Lect. XXVIII, p. 80.

periments and their correct evaluation[239], a critical examination that is necessary when a new research strategy is set up in order to limit the scope and contents of the research (Subsec. 3.3.3). In conclusion, it is possible that what became beneficial for science in later stages was less beneficial to continental philosophy and theology that remained trapped in an epistemologically less fruitful attitude (Subsec. 2.3.2).

Now, coming back to Hegel and his dialectic, we may consider here a further fundamental difference with respect to our approach. The three philosophical principles that we have formulated in this chapter are not only heuristic principles but are addressed to the *natural* world. Peirce also elaborated his three categories as addressing a basic understanding of our universe[240]. On the contrary, Hegel's dialectic categories present themselves as concerned with *being* as such, and even as expressing the quintessence of the Absolute. In the foreword to the first edition of his *Science of Logic* he says that the logical science exposed therein constitutes the true metaphysics or the pure speculative philosophy[241]. Moreover, this Absolute is understood as a mind-like substance. This is a huge step, since it resolves the problem of the universe and even the problem of being in the activity of mind. When we speak of an Absolute or of Being as such, we are naturally led to theological issues. Indeed, we hope that our principles may also show some connection with metaphysics and theology (we shall explore some of these possibilities in the next chapter). However, we are aware that any possible connection in this sense should be articulated by explicitly taking into account the developments in theology and interacting with theologians. Instead, we feel that a significant part of the Hegelian and post-Hegelian continental philosophy has developed a sort of abstract rational theology that is, however, deprived of those contents and elements that are relevant in the true rational theology, since they are connected with Revelation or Creation (Subsec. 2.3.7). This kind of philosophy becomes a sort of theology of Being without God and even without universe. In the words of (the so-called *second*) Heidegger: "Thought, said directly, is the thought of Being. The

---

[239] See the insightful analysis provided in Harrison 2007, pp. 172-185 and 198-216. Philosphers like Locke (1689) preferred to develop an inquiry on the limits and scopes of the human mind, an investigation which was also part of the orginal research program of Bacon. In doing so, however, Locke gave a psychological but nevertheless objective turn to Descartes' introspective approach. We find this approach as very helpful for dealing appropriately with the basic idealistic insights.

[240] Peirce *CP*, 1.162, 1.175 1.322, 1.351.

[241] Hegel 1833, v. 1, p. 5.

genitive is twofold: Thought is of Being as far as thought happens from Being, belongs to Being. Thought is also Thought of Being, inasmuch as, belonging to Being, thought listens to Being"[242]. This approach is characterized by two main misunderstandings: on the one hand, an onto-theology without theology and, on the other, the claim to develop an ontology without any reference to the empirical contents (Subsec. 2.2.3). The consequence could be a distortion of philosophy (or metaphysics), since the critical (and questioning) methodology of the latter does not endure a too strict connection with the theological methodology of listening. Philosophy needs to find a correct relation to theology, being able of simultaneously acknowledging what comes from the latter discipline, which cannot be found by philosophical means alone, and remaining independent in its investigation. Nevertheless, in the mentioned Heidegger's approach there is also an important insight that we shall develop later on.

One of the main problems that we still have today is that many philosophers consider these developments to be in perfect continuity and coherence with the previous philosophical tradition. Actually, some could regard this approach as an ulterior (and even conclusive) step relative to Middle-Age and Modern philosophy in further developing a doctrinal body of knowledge that ultimately stems from Aristotle and neo-Platonism. Actually, it is perfectly true that Aristotle speaks of *Metaphysics* as dealing with being as such and the attributes which necessarily belong to being[243]. The point is to understand the way in which he conceived being as such, as the main object of what he called "first philosophy" (see also Subsec. 2.2.7). After showing the four senses (being in the accidental sense, being according to the categories, being as true and false, and finally being as potentiality and actuality[244]) in which the term "being" may be used (or the four *modes* of being that can be acknowledged, as Aquinas puts it[245]), he suggests that these senses share a common reference, that is, they all refer to some kind of *substance* (Subsec. 3.1.8).[246] Therefore, first philosophy should ultimately address substances in their specificity as it is explicitly pointed out in the twelfth book of the work (Λ). St Thomas supports this view in his Commentary. First, he points out that any mode

---

[242] Heidegger 1946, p. 316. See also Heidegger 1949; 1954.

[243] Aristotle *Met.*, 1003 a 20.

[244] Aristotle *Met.*, 1017 a 7-b 9.

[245] Aquinas *In Met.*, a. 540.

[246] Wedin 2000.

of being (quality and quantity, motion and generation, or negation and privation) is called "being" inasmuch as it has relationships with an ontologically primary thing as *existing by itself* (in its separated individuality), i.e. a substance. Accordingly, he states that "all other modes of beings are reduced to this as the primary and principal mode of being"[247]. Thus, he clearly maintains that first philosophy is, according to Aristotle, chiefly concerned with the study of substances[248]. Moreover, he points out that first philosophy should take into account *all* substances as to their being, while each of them, when considered as being a particular *kind* of substance, should be the object of a special science[249]: all of Physics (covering at that time all of natural sciences) should deal with things that are individually separated but changing, mathematics (likely covering all of formal sciences) with things that are not individually separated (not substantial) but changeless. Metaphysics also deals with particular kinds of objects, namely substances that are changeless[250]. Therefore, Aristotle argues that: (1) the object of this enquiry should be represented by those things that are separable from matter, immobile and therefore eternal and (2) these are the principles and causes of the *beings* as beings or the common *causes* of the sensible things that are visible to us[251]. As to point (1), it seems that Aristotle is actually taking transcendent reality into account, so that we are directly dealing with theological matters according to the terminology and conceptions that the author himself adopts. Indeed, he states that "it is obvious that, if the divine exists anywhere, it exists in this kind of nature"[252] (i.e. eternal, immobile and separated from matter), and, provided that first philosophy is addressed to this object, it is to be called theology. Point (2) raises a very huge problem. How can we understand the causal relation between a transcendental substance and the sensible ones? St Thomas says that "we must remember that even though things which are separated from matter and motion in being and, in their intelligible structure, belong to the study of first philosophy, still the philosopher not only investigates them but *also sen-*

---

[247] Aquinas *In Met.*, a. 543.

[248] Aquinas *In Met.*, a. 546. Also Averroes, *AM*, p. 25 (j32), says that the two main parts of Aristotle's *Metaphysics* are devoted to sensible (material) things and separated substances in their relation to God, respectively.

[249] Aquinas *In Met.*, a. 547.

[250] Patzig 1960. See also Wedin 2000; 2009.

[251] Aristotle *Met.*, 1025b1-1026a32.

[252] Aristotle *Met.*, 1070a31-1072a18

sible things inasmuch as they are beings"[253]. Therefore, it seems that, on the one hand, first philosophy is still linked to philosophy of nature as it also concerns sensible substances as such and their necessary (not accidental) attributes[254]. On the other hand, first philosophy is committed to theology since it also encompasses the enquiry about those substances that are not sensible, not dependent on space-time and matter constraints, and yet have a causal influence on the sensible substances. We may recall here that this was one of the main criticisms Aristotle addressed to Plato's theory of forms[255].

We are fully aware that there is a huge amount of different interpretations of these fundamental paragraphs. However, we think that a viable and fruitful interpretation of the way Aristotle understands the relation between intelligible substances, especially the primary motor, and the sensible ones is the following: it is doubtful that we can identify his understanding of God as the principle of motion with God as first or eminent being (as it is customary in the Christian tradition).[256] As a matter of fact, in the examination of God as the principle of motion, Aristotle emphasizes that such a principle (i) is necessary, (ii) is the good, (iii) cannot be different from how it is, (iv) its mode of life is excellent, (v) is eternal, (vi) its actuality is also pleasure, (vii) is thought of thought since it is self-sufficient, (viii) is life, (ix) motionless, (x) separated from other things, (xi) without parts and indivisible, (xii) without passions and inalterable[257]. Although showing a remarkable level of reflexive speculation, some of these thoughts have an evident root in pagan religion and mythology. It is Aristotle himself who says that the essence of pagan religion, that is, the statement that the first substances are gods, is not only true but even the result of a divine inspiration, although he adds that such an insight is also encrusted with mythical contaminants[258], which is fully in accordance with our understanding of ancient philosophy as a purification from mythology (Sec. 1.2).

---

[253] Aquinas *In Met.*, a. 1165 (author's emphasis).

[254] The point is also supported by the extreme attention Aristotle draws throughout the book H (VIII, 1042a 3-1045b 23) to the nature of sensible substances. Aquinas turns back to the problem in his comments on Aristotle's book Λ (XII), stating that first philosophy "considers both sensible substances and immovable substances inasmuch as both are beings and substances" (*In Met.*, a 2427).

[255] Aristotle *Met.*, 991a8-b9.

[256] Taylor 1998.

[257] Aristotle *Met.*, 1072 b 9-1073a14.

[258] Aristotle *Met.*, 1074 b 1-14.

We may also add that Aristotle has absorbed from the theological purification of pagan religion the idea of God as autarchic, detached and inalterable, which is also shared by Plato's world of forms.

What we are trying to say is that, in order to consider God as the eminent being and to develop an ontological relation between God and the other beings in terms of a participation or even a gradation, one needs to have already developed a theology of Creation[259]. It is true that St Thomas assumes that Aristotle's first philosophy ultimately deals with causes of the being as being because the concerned things "are being in the highest degree"[260], thus suggesting that the causal relation existing between the transcendent substance and the sensible ones is here a matter of ontological primariness and participation. But obviously this must be read in a context in which medieval scholars falsely attributed to Aristotle some texts that were in fact written much later by neo-Platonists, like the *Theology of Aristotle* and the renowned *Liber de causis*, constituting, together with other texts, the corpus of the so-called *Plotiniana Arabica*[261]. This raises the complicated problem of the role of neo-Platonism in the reception of Aristotle by Latin scholars of the 13th century. Ordinarily, we understand today Plotinus and neo-Platonism as representing the speculative bridge between Aristotle and Plato, on the one hand, and Middle-Age philosophers and theologians on the other. This is clearly true since the fundamental conceptual nexus that we need here in order to build an *ontological* understanding of the causal relation between God and finite beings is an idea of emanation of the latter from the first principle, which both transcends them and is above any limitation characterizing finitude. Indeed, only in this way we may understand God not only as the cause of specific effects (like motion) but of the *being* of everything else (although there are obvious significant differences between Plotinus' idea of the One and St Thomas' conception of God as *ipsum esse per se subsistens*).[262] However, one should consider that Plotinus lived in the 3rd

---

[259] Therefore, we agree with the difficulties that Wedin (2009) raises about the way Aristotle considers the relation between God and finite substances: in other words, the great philosopher is not able to fully develop his own insight.

[260] Aquinas *In Met.*, a. 1164. Indeed, St Thomas here also states that for Aristotle God is the cause of the substance of all other things and even of the substance of heavens, not only of their motion, contrarily to what Averroes thought.

[261] Gilson 1944, pp. 345 and 377-378. It is likely that St Thomas himself did not read some of these texts directly but was nevertherless indirectly influenced (Taylor 1998).

[262] Taylor 1998.

century, in a cultural melting pot in which not only the revealed religions (the Judaic and Christian ones) were widespread, but also an important theological thought was already developed by eminent representatives of both religions. Moreover, many Gnostic sects introduced distorted ideas about different divine hypostases in a progressive although dramatic (and still mythological) process of emanation. As is well known, Plotinus himself devoted a part of his treatise to a sharp criticism of Gnosis[263]. It is not by chance that E. Gilson emphasized that the philosophy of Plotinus cannot be understood as a deductive development of Plato's one but as aimed at solving a new problem, that is, the unification of a philosophical and religious order of consideration[264].

In conclusion, we think that any deep metaphysical thinking about the first cause of beings (like those of Aristotle, Plotinus or St. Thomas) always presupposes the existence of both religious traditions and their de-mythologization through theology (the pagan religion for Aristotle, the Hellenistic religious melting pot for Plotinus, and the Christian tradition for St. Thomas). Therefore, a true metaphysics dealing with the ontological question of being, or of the first Being, cannot be developed without an appropriate theological background: this ontological relation between the first Being and finite beings can be understood today only by taking into account a theology of Creation. We share here again Gilson's judgment that this approach is necessarily an ontology entirely oriented towards a theology[265]. It is time to overcome the misunderstanding that philosophy can have a material object as distinct from those of the empirical sciences on the one hand and that of theology on the other (Subsec. 2.2.3). However, this does not prevent philosophy from having a well-defined *formal* object, as we have shown in the present chapter and summarized in Subsecs. 2.2.6-2.2.8. In other words, philosophy and its different branches cannot be self-grounded and rationally justified without taking into account other domains of experience and knowledge. Therefore, our ideal of philosophy is to a certain extent of a Kantian type[266]: it is a formal and architectonic discipline aiming in particular at showing limitations

---

[263] Plotinus *Enn.*, II ix. See also Katz 1954.

[264] Gilson 1987, p. 41. According to Draper (1875, pp. 122-123) also an oriental, Vedic component enters into this mix, especially when we consider the re-absorption of the human soul in the One.

[265] Gilson 1987, p. 87.

[266] Kant 1787.

and incongruence in any cognitive field thus providing conceptual clearness and cleanness (see also Subsec. 2.3.7). This is not detrimental to philosophy's autonomy. The correct point of view is to conceive philosophy not as a theory or a conglomerate of theories, or even as a system organizing specific contents. Philosophy is rather an activity, which we have stressed to be a critical one. In this critical activity, philosophy is not only autonomous from, but even sovereign with respect to, the other two cognitive fields since its business is to critically evaluate their statements. On these presuppositions, it is now time to pass to the second aspect of the problem of this book, that is, the contribution of theology to this reflection and the relations that can be established with philosophy.

4

THEOLOGY OF NATURE: A STARTING POINT

Whilst theology shares with science and philosophy the study of the universe and man, which it enquires with its specific methodologies, it also has its own material object, namely God and the religious traditions. All the reports on original experiences of Revelation can be considered, from a phenomenological point of view, to be part of cultural and religious traditions (Subsec. 2.3.1). For this reason an important concern of Christian theology is the interpretation of the Sacred Scriptures in a congruent way with respect to the knowledge that mankind strives to achieve studying the created natural world, as already understood by St. Augustine (Subsec. 2.3.2). When modern science established its methodology and consolidated its results, thus playing a more and more relevant role in Western culture, part of this task was to show that Scriptures do not contradict scientific results; remarkable efforts have been made accordingly also in recent times, as we have already mentioned (Sec. 1.1). As to this matter, we note that scientific advancements, from time to time, put into question what is assumed to be the "traditional interpretation" of Scripture. Indeed, the Bible was written in a particular historical and social context[1], which represents a real challenge for theologians, and consequently a critical comparison with reference to the *actual and current* state of human culture is definitively needed[2].

The fact that the culture of our time is dominated by science suggests that a theology interested in developing a fruitful understanding of the natural world, as well as of the current status of our culture, needs to develop a rational enquiry thanks to the interaction with natural sciences, a step further with respect to a simple non-belligerence agreement[3]. This constructive dialogue between science and theology should take into account the supply offered by philosophy of nature, with

---

[1] John Paul II 1981. See also Maldamé 2003, pp. 140-143.
[2] Poupard 1983; 1992.
[3] Haught (2000, pp. 31-36) analyzes "separatist" attempts at keeping theology apart from current evolutionary theory.

the goal of finding consonance and fruitful relations between the two[4]. We recall that traditional Medieval theology also took recourse to similar means when dealing with philosophical and theological issues of general interest as, for instance, in the case of the argumentations that St Thomas addressed to non-Christians[5].

In the previous chapter, we have outlined the principles and some of the main contents characterizing the philosophy of nature that we have proposed as the mediating instance in this dialogue. It is now time to draw our attention to some issues of that branch of theology that we are proposing as one pole of the dialogue (the other being science), that is, to outline principles and contents of what we have called "rational theology of created nature" or, briefly, "theology of nature". We stress that whilst we have developed a relatively articulated philosophy of nature, our aim here is only to provide some general hints for a theology of nature. Starting from our philosophical examination, we wish to suggest some conceptual developments to professional theologians, in order to explore (with their help) whether or not we can establish an interesting field of interaction. We recall that the term "theology of nature" indicates that we are looking for a theological framework compatible with the proposed conception of naturalism (Subsec. 2.2.5) which, as mentioned, is not confined to, neither should be identified with, reductionism or materialism. We are aware that some influential interpretations of naturalism have induced many Christians to express significant concerns about the use of this term[6]. However, we hope that our previous proposal may help to restore a more positive attitude that could also turn out to be helpful in the dialogue with non-believer scientists.

In proposing the following outlines of a theological enquiry, it is important to avoid generalizing specific scientific results or concepts (Subsecs. 2.2.3, 2.3.3 and 2.3.4). Throughtout the history of theology, there have been many of such hypergeneralizations. Some examples are: the concept of absolute Space and Time, introduced by Newton who understood them as God's sensoria[7]; Darwin's theory of evolution, very soon incorporated by the British theology of incarnation (and by Teilhard de Chardin in continental Europe) for justifying the arising of a new

---

[4] See also Pannenberg 2008, pp. 27-30.

[5] Aquinas *Cont. Gent.*

[6] For example, D. Alexander and R.S. White maintain that Christians have no need to attach the connotation of naturalism to their views, if they are aiming at describing the world created and actively sustained by God (Alexander–White 2004, pp. 29-31).

[7] Newton 1697, Scholium generale; 1704, pp. 403-404; Koyré 1957, Chs. 7-10; Pannenberg 1993, pp. 59-65.

humanity[8]; the concept of field theologically interpreted by Torrance and Pannenberg (although with some caveats) as a bridge to the spiritual dimension[9]; the cosmological theory of the Big Bang understood as supporting the doctrine of Creation[10]; the countless number of philosophical (and theological) abuses of quantum-mechanical concepts supposed to ground or justify any sort of speculation. The problem is that scientific conclusions are overcome with time. Moreover, being expressions of specific theories, they may lead to several philosophical and theological distortions and overtones when taken out of their empirical context. It is therefore crucial to start from general philosophical lessons drawn from science and to make use of specific results only as supporting those general tenets.

4.1 A Look at St. Thomas' Approach

It can be affirmed[11] that there are three ways in which the Christian tradition tries to cope with the relation between God and creatures:

(1) Following St Augustine in finding out the seeds of divine wisdom in the fabric of the cosmos.
(2) The methodology of *analogia entis*.
(3) The Barthian *analogia fidei*, in which the connection between God and creation is established by Him through His self-revelation in the Scripture.

We shall not directly enter into issues of Biblical hermeneutics here, which we leave to authors who are more competent than us; therefore, we will not deal with the last point. We have also already introduced some sparse considerations that are related with approach (1), especially in the context of our examination of the issue of intelligibility. In this chapter, we shall focus on point (2) and in particular examine one of the most fundamental contributions of Christianity under this respect, that is, Saint Thomas Aquinas' theology of Creation[12]. We shall address the questions related to the procession of the finite creatures from God, as this con-

---

[8] See Pannenberg 1993, pp. 56-58 and 47.
[9] Torrance 1969; Pannenberg 2008, pp. 37, 45, 64-72; Pannenberg 1993, pp. 37-41.
[10] See Coyne 1998.
[11] McGrath 2001-2003, v.2, pp. 64-78.
[12] The Theology of Creation is not only a helpful background for the dialogue with science but also a very important common ground for interaction with other monotheistic religions: see for instance Guessoum 2011, p. 46.

stitutes the main theological background of theology of nature. This means that we consider the relation between God and nature always from the perspective of the latter as stressed in Subsec. 2.3.6. Furthermore, we do not aim at formulating a commentary on St. Thomas' work but simply to give a short account of some of the main ideas that he developed in this field that turn out to be very helpful for framing our approach.

4.1.1 *Intelligibility and God's Mind*

According to St Thomas, the world, in any of its manifestations, is grounded in God's mind, that is, in the ultimate principle of intelligibility[13] (Subsec. 3.2.6: see the 7th question in Subsec. 2.3.7). In other words, the causal procession of the whole being from its universal Cause is considered by St Thomas as an *intelligible* emanation[14]. This is one of the most difficult issues across the history of Middle-Age theology and also touches on Modern-Time speculation. Since the world does not come about by chance (Subsec. 3.3.6), there must be an idea or reason for the universal order that we experience[15]. St Thomas affirms that it is God Himself and His divine wisdom that is the exemplar of everything else[16]. As the great commentator of St Thomas and expert of Middle-Age philosophy E. Gilson says[17], we must assume that these ideas exist in God without introducing any multiplicity, so that they are in reality only one intelligence. This claim is connected with St Thomas' view that God's wisdom is not discursive like the human one[18]. It is only in their intelligible manifestations and not in themselves that God's ideas undergo a distinction and become distributed according to a specific order[19]. The theological reason is that the original idea itself is the *Logos*[20], who is also the original model of the whole

---

[13] This is also the case according to the Jewish tradition: Maimonides (*GE*, I, § 72) says that the relation that God has with the created world is comparable to that of the human mind with man.

[14] Aquinas *S.Th.*, I, q. 27 a. 1.

[15] Aquinas *S.Th.*, I, q. 15 a. 1-2.

[16] Aquinas *S.Th.*, I, q. 44 a. 3.

[17] Gilson 1944, pp. 210-211 and 534-535.

[18] Aquinas *S.Th.*, I, q. 14 a. 7. This is why Cajetan (1514) in his commentary to I, q. 15, a. 1, says: "nihil enim Deus intelligit extra se, sed omnia in se tantum". See also De Morgan 1847, pp. 44-45.

[19] In his commentary (1514) to I, q. 15, a. 2. Cajetan distinguishes between ideas' multiplicity *quod* (different ideas referred to different things) and *quo* (when we have different ideas of the same thing) and adds that only the second one contradicts God's simplicity.

[20] Tanzella-Nitti 2009, pp. 127-128.

of creation and reason for everything, and so is undivided in Himself[21]. In this way, we can avoid assuming a multiplicity of ideas that is co-eternal with God as Scot Erigena did. Therefore, as far as God knows Himself, not only as He is in Himself but also as the Being to which creatures participate (Sec. 3.4), He generates forms as a certain possible participation of a particular creature to His perfection: according to St Thomas, what is one and simple in God is multiple and divided in finite beings[22]. This is why St Thomas says that those forms instantiated in the finite world display some similitude with the absolute unity of God's wisdom, but also that the world can be said to have been created to the similitude of the intelligible forms that derive from God's mind[23]. This intelligible emanation is what we understand as Creation.

The fundamental implication of this doctrine is that the variety of happenings in our world that we must consider to be contingent, by chance, and in a way irreducible to the intelligible structure of the world (at least as far as we can understand it), nevertheless remain grounded in the ultimate principle of *any* intelligibility: God's mind[24]. This theologically grounds a rational faith in the possibility of providing for explanations that all scientists and philosophers seem to show even when facts apparently contradict their expectations (Subsec. 2.2.8), without incurring in the error of rejecting (as Einstein did, Subsec. 2.3.5) the possibility of random happenings in our world since they could give raise to a conflict with our (limited) understanding of intelligibility. So, theology opens a fruitful "cognitive gap" between God's mind and our lawful explanations.

4.1.2 *Creation and Creatures: Primary and Secondary Causes*

The theory of causality in St Thomas, especially the distinction between God as *primary* universal Cause[25] and the finite *secondary* causes, is something that allows for a deeper understanding of the happenings in our world and of the way in which the latter can be considered as *intelligible*. It is well known that the tra-

---

[21] With the words of Maimonides (*GE*, I, § 68): God and the object thought of or perceived by Him are the same thing.

[22] Aquinas *S. Th.*, I, q. 47 a. 1.

[23] Aquinas *S. Th.*, I, q. 15 a. 1; q. 44 a. 3. And therefore it is legitimate to conclude in the way Cajetan (1514) comments on I, q. 14, a. 13: "dicitur quod [...] ideae sint ipsa essentia Dei ut est obiectiva rerum ratio, optime dictum est quod, quia rationes seu ideas rerum praesentes apud se habet Deus".

[24] Lonergan 1957, p. 668.

[25] Aquinas *S. Th.*, I, q. 4 a.3; q. 44 a.2; q. 45 a.1. See Pangallo 1991.

ditional interpretation of Aristotle[26] developed in the Islamic world (that also became canonical for the initial Christian understanding of these matters), considered God as being the first term (the non-moved motor) of a causal chain determining the motion of the whole universe through a gradation of beings (Sec. 3.4). For the ripe age of scholasticism, things seem to stand in a different way[27]. According to St Thomas, whilst the finite causes simply bring about specific effects, the action of God is *Creation*. It is crucial to understand that Creation is something that involves much more than just the beginning of the universe as a sort of initial input[28] (Subsec. 2.3.3). Indeed, God's action is rather a continual Creation (*Creatio continua*).[29] Such a term means that God is the cause of the original constitution of the being from every viewpoint[30], which also implies that we cannot assume any form of model whatsoever pre-existing the existence of finite beings[31] apart from God's mind. In other words, God is the cause of both the *existence* and the *conservation* of any finite reality, including their forms[32], that is the way in which we can theologically deal with the first question raised in Subsec. 2.3.7. The Jewish theologian F. Rosenzweig stressed the fundamental transience of finite beings by using the word "inessential being" (*unwesenhaftes Wesen*).[33] In that context, Rosenzweig recalled the Jewish thanksgiving-prayer of the morning that praises God for having rebuilt the creation again and again. It is clear that this doctrine evokes the radical dependence of creation on God.

As the source of any being, God also *empowers* any creature with a dynamism able to produce particular effects (i.e. to be a finite cause). This is known as the doctrine of the act of being (*actus essendi*), that is, the constitution of the existence of any individual creature as the original source of further effects[34]. God also remains the cause of the latter, but only indirectly through the modality of causation

---

[26] Todisco 1999.

[27] Aquinas *S. Th.*, I, q. 45 a. 3; *De Pot.*, q. 3 a. 3.

[28] Aquinas *S. Th.*, I, q. 46; Maldamè 2010; Pannenberg 2008, p. 35; Polkinghorne-Beale 2009; Alexander 2008, Ch 2.

[29] Aquinas *De Pot.*, q. 5, a. 1.

[30] Aquinas *S. Th.*, I, q. 45, a. 1, 2, 5; *De Pot.*, q. 3 a. 3; *Con. Gent.*, b. II, ch. 6, 15, 26; *Sup. Sent.* b. II, d. 1, q. 1 aa. 1-2.

[31] Aquinas *S. Th.*, I, q. 44, a.1.

[32] Aquinas *S. Th.*, I, qq. 103-119.

[33] Rosenzweig 1921, §§ 103-104.

[34] Aquinas *S. Th.*, I, q. 3 a. 4; *In I Sent.*, d. 8, q. 5, a.2; *Con. Gent.*, b. II ch. 18, 21; *De Ver.*, q. 1, a.1. See also Pangallo 1987

176

of those finite beings. This crucial difference between the act of being, and therefore the existence of any creature on the one hand, and the finite effects that the latter may bring about, on the other, helps us to understand that the act of being is communicated to any finite being *directly* and continuously[35], and not in a sort of cascade-like process as claimed by Avicenna's interpretation of Aristotle[36] which was still encrusted with the idea of the First Motor (Sec. 3.4) : only God is Being as such (*ipsum esse*), whilst to all creatures being can be ascribed in a subordinate way, as far as they are all supported by the direct creative action of God. This implies the absolute sovereignty of God on everything else and the impossibility to know Him or to define Him or even to prove Him if not through some effects[37]; moreover, this also implies that God's causality does not interfere at all with the way in which finite causes work and neither does violence to any natural course of things[38], whilst it is a *real relation* that any being has with its Principle[39].

---

[35] Aquinas *De Pot.*, q. 5, a. 2; see also Maréchal 1926, p. 181.

[36] Avicenna *GR*, I, Secs. XXXVIII-LI ; II, Sec. IV. We recall that Avicenna (*GR*, I, Sec. XXX-XXXVI) assumed that the world is eternal since it derives from God's continuous action. Here, God acts on contingent creatures through the intermediation of celestial spheres. A similar statement can be found in Averroes, *AM*, pp. 23-24 (j31-j32), when he distinguishes between "that which moves" and "the agent". However, Averroes' position is more refined since he (against Avicenna) says that the former only gives its motion to what is moved, whereas the agent gives the form through which the motion occurs. In this way, intelligences moving the celestial bodies (according to efficient, formal, and final causation), transmit the forms to the sensible universe, as stated in Averroes, *AM*, pp. 151-153 (j148-j149). For this reason, in *AM*, 156-157 (j152), he compares the disposition of the world with that of the city-state ruled by an aristocracy, whose leadership is multiple but forms one leadership directed towards one end. On this subject see also Th. Aquinas *De Pot.*, q. 3, a. 3. It is interesting to observe that J. Draper (1875, Chs. 4-5), supporting the traditional emanationism against Aquinas' reform of the concept and in particular the idea of the universal *Intellectus agens*, underestimates the historical relevance of the Thomistic assimilation of Aristotelian philosophy, even for the birth of modern science.

[37] Aquinas *S. Th.*, I, q. 3, a. 5. Similarly, Maimonides (*GE*, I, § 57) says that God's existence is His essence. For this reason, Maimonides (*GE*, I, §§ 52 and 56) says that existence can be attributed to God and the creatures only in a homonymous way (St Thomas would say in an analogical way). For the above reasons, in the same article St Thomas refuses the idea that God can be included in any genus. The same point is stressed by Maimonides *GE*, I, §§ 50-51 and 58.

[38] Maldamé 2003, pp. 241-243. Tanzella-Nitti 2009, p. 45. Edwards 2010, pp. 62-64 and 80-84.

[39] Aquinas *S. Th.*, I, q. 45, a.3; *De Pot.*, q. 3, a. 3 and q. 7, a. 8. In his comment (1514) on the quoted article of the *Summa*, Cajetan says that "Creatio ponit in creato relationem tantum". See also Salucci 2011, § 6.4. This may help us to overcome Einstein's fears that a personal God

Our brief account gives an idea of how insightful theology may be in dealing with scientific matters (Subsec. 2.3.5).[40] St Thomas' concept of Creation (and of the act of being in particular) allows us to seriously take into account and simultaneously understand the ontological gap that exists between formal constructs and existing beings[41]. Even supposing that we could have a perfect computer-like description of our universe, this would not mean at all that this description *is* our universe (Subsec. 3.3.4). The reason is that any description, law, theory, mathematical formulation or structure is not provided with the *power* to produce particular effects[42]. The great philosopher Leibniz said that ideas do not act[43] (*Ideae non agunt*).

Considered from this perspective, the problem of a potential conflict between the theory of evolution and faith in a Creator can be solved at the start[44]: the way in which the finite causes act on each other in no way determines or diminishes God's causation, which deals with being as such, and with the created world's existence in particular (Subsec. 3.3.6). We also mention that such a theology helps us to avoid mistakes like the sort of "perpetual creation" introduced as an *ad hoc* hypothesis for preserving the steady-state of the universe (Subsec. 2.1.6).

4.1.3 *Existence and Contingency*

According to St Thomas, human knowledge can only be knowledge of universals, and never of singulars[45], while God's knowledge can also be of singulars[46]. In the

---

may interfere with finite causes by way of His action, as also remarked by a fine theologian like P. Tillich (Jammer 1999, pp. 80, 94-95, and 108-109). We shall come back to the issue of the personal God at the end of this chapter.

[40] Benedict XVI 2011.

[41] Aquinas *S. Th.*, I, q. 45 a. 4.

[42] Aquinas *S. Th.*, I, q. 105 a. 5.

[43] Leibniz 1678-1679, p. 150.

[44] Maldamé 2011; Haught 2006, p. 16.

[45] Aquinas *De Ver.*, q. 2 a. 6; Lonergan 1957, p. 54.

[46] Aquinas *De Ver.*, q. 2 a. 5, 7. This seems to be a major difference relative to Avicenna (*GR*, I, Sec. XIII) who maintaned that the necessary existent being cannot conceive finite beings as changeable and corruptible, since this would somehow make itself changeable and corruptible. Similarly in Averroes, *AM*, p. 158 (j154). But also the intelligences ruling celestial bodies do not conceptualize beings below them. However, since through a gradation of intelligences we finally reach the human intelligence that is able to think of sensible beings, this is the way in which an intelligence of these beings is propagated from God down to finite beings, as explained in Averroes, *AM*, pp. 158-161 (j154-j157).

previous chapter, we pointed out that natural science is more and more understood as something dealing with general predictions, and not with singular happenings as such (Subsecs. 3.2.3 and 3.3.8), although we can have direct experiences of these singular happenings, at least in a negative way (introduction to Sec. 3.2), so that they constitute an integral part of the body of our knowledge. It is in this sense that our principle of dynamicity (Subsec. 3.3.3) dynamically connects random events with the source of any intelligibility in our world: formal constraints, which ultimately stem from God's mind (Subsec. 4.1.1). In the long question 14 of the first part of the *Summa Theologiae*, St Thomas[47] distinguishes between God's science about things that never occur, which he calls "simple intelligence" (*simplex intelligentia*), and God's knowledge of things occurring in nature, which he calls "vision" (*scientia visionis*). In such a context, St Thomas stresses that God's modalities of knowing individual and contingent happenings is a-temporal, since they are *immediately* present to Him as the Being outside of the temporal course of our universe[48]. Given the distinction between primary Cause and secondary causes, such a knowledge of individuals does not abrogate their irreducible alterity.

This helps us to understand an important consequence of St Thomas' conception of Creation: the possibility that happenings are *contingent* even if they *depend* on God's government[49]. Indeed, St Thomas in the long article devoted to the issue of the so-called contingent futures[50] admits, following Aristotle, that there can be contingent realities that are not predetermined by some finite causes (Subsec. 3.3.4). Following Rosenzweig, we could even say that the irreducibility of happenings is the surest mark of creation[51]. So, St Thomas acknowledges the occurring of

---

[47] Aquinas *S. Th.*, I, q. 14, a. 9. Cajetan (1514) comments that vision supposes the real presence of the object.

[48] Aquinas *S. Th.*, I, q. 14 a. 13. See also Lonergan 1957, pp. 402-404. The issue of how contingent happenings that have not yet occurred (the so-called contingent futures) could be present to God, is one of the most difficult problems. In his insightful commentary, Cajetan (1514), taking into account Duns Scot's objections, says that we should distinguish between a *nunc temporis* and a *nunc aeternitatis* and adds that only in this second sense it can be said that contingent futures are present to God's eternity.

[49] Peacocke 1993, pp. 115-121. See also Edwards 2010, pp. 52-55.

[50] Aquinas *S. Th.*, I, q. 14 a. 13. See also Aquinas *De Ver.*, q. 2 a. 12.

[51] Rosenzweig 1921, § 118. Moreover, according to Pannenberg (1993, p. 76), theology understands reality primarily by contingency of occurrences "that take place constantly" and "that are experienced as the work of almighty God". This is the main reason why Pannenberg is convinced of the importance of theology in order in order to fully understand nature (p. 48).

chance components in some finite beings, although these beings are still under God's government[52]. Here, we are in the presence of a rationally plausible way to intend the natural created world, while avoiding extreme forms of both determinism and indeterminism possibly suggested by unilateral philosophical interpretations of scientific knowledge. On the one hand, only a radical indeterminism denying any form of ontological determination and law would be in opposition to this Aristotelian-Thomistic conception of nature[53] (Subsecs. 2.2.5 and 3.3.7) but *not* the kind of indeterminism suggested today by science and expressed by our three principles of philosophy of nature[54]. On the other hand, St Thomas rejects the idea of a necessary causal chain in our world[55], thus providing the key to overcome the conflict between our experience of contingency and the conception of God's action as a trigger of a causal chain, which Duns Scot identified in Aristotle's philosophy[56], at least as far as it was understood through its reception in the Islamic world (the so-called *Plotiniana Arabica* that influenced philosophers like Avicenna, Sec. 3.4).[57] It is worth mentioning, however, that Aristotle's original position is probably more complex according to his discussion about future contingent happenings[58]. This examination shows how fruitful is the Thomistic theology for dealing with the second and fourth questions raised in Subsec. 2.3.7.

This is related to the way in which many people still misunderstand St Thomas' fifth way of proving the existence of God[59]. For a correct understanding of such an argument we need to consider that the distinction between primary Cause and secondary causes does not only apply to efficient causation but also to final causality[60]. Finite final causes, precisely as it happens for efficient causes, are subordinated

---

[52] Aquinas *S. Th.*, I, q. 103 a. 5; q. 116 a. 1-2.

[53] Selvaggi, 1964, pp. 147-48 and 157-67. Tanzella-Nitti 1997 pp. 15-16.

[54] On this point see Selvaggi 1985, pp. 417-419 and 525-26.

[55] Aquinas *S. Th.*, I, q. 106; q. 115; q. 116 a. 1-4.

[56] Aristotle *Phys.* 259b32-260a19; *De caelo* 286a34-286b9. Duns Scot *Ord.*, I, d. 2, p. 1, q. 2, n.87. It is worth noting that Duns Scot strongly stressed the ever-changing character of material beings and therefore the intrinsic temporal dimension of any contingent reality, implying its radical incompleteness (Duns Scot *Ord.* I, d. 39). See also Pedersen 2007, pp. 190-193; Salucci 2011, § 5.17.

[57] Taylor 1998.

[58] Taylor 1957 and Hintikka 1964.

[59] Aquinas *S. Th.*, I, q. 2, a. 3.

[60] Here, the source of the problem seems to be the statement of Averroes, *AM*, p. 23 (j30-j31),

to God's Providence and Government but cannot be identified with the latter. This is the reason why St Thomas, in exposing his fifth way, stresses that this proof deals with God's government of all things (*ex gubernatione rerum*): since we have just seen that this government also applies to things that happen by chance (which by definition are not provided with finality), it seems obvious that St Thomas is not saying that the whole world is ruled by an intrinsic finality, like some stoic approaches maintained, or that God's providential action and creatures' finality are strictly intertwined, as Leibniz maintained (Subsecs. 2.3.3 and 3.3.6). Since both the processes happening by chance and those endowed with finality must depend on God's government, St Thomas' fifth way cannot be considered as an argument from design. This means that such a proof has a *metaphysical* character[61] without any connection with the specific final causes involved in particular happenings and contingencies. The essence of the proof, instead, is that any being, in its participation to the absolute Being, is part of a universal order intended by God. For this reason, according to Cardinal C. Ruini[62], St Thomas' argument in the fifth way is strictly connected with the intelligibility of our universe rather than with the finality displayed in some particular natural processes or even in the whole universe. Thus, it is unjustified and naïve to connect these two different orders of questions and to let the actions or dynamics of finite beings be endowed with a special, supernatural significance. A philosophical appraisal of the Darwinian theory of evolution shows the impossibility of considering the *entire* universe as being teleological in character, helping to disentangle the biological domain from supernatural causal factors (a philosophical-theological no-go theorem based on the first principle above, Subsec. 3.1.8). It is again a philosophical reflection on modern biological knowledge that may appropriately reassess teleological causation within the biological domain, attributing it to the *individual* organisms' activity during their lifespan and not to

---

that in the physical world we deal with material and efficient causes (a statement that would be agreed by many scientists) whilst when dealing with the divinity we need to consider final and formal causation. Nevertheless, Avicenna's and Averroes' reflections on God's causality remain an important presupposition of the Thomistic synthesis. To understand how difficult this problem is, one needs to consider the philosophical school of Al-As'ari that supported the idea that no causality at all exists in our world except for the sole divine power to create transient configurations of things moment by moment according to His will (see Frank 1966; Vadet 1976).

[61] Gilson 1960.

[62] Ruini 2010.

their evolution[63] (Subsec. 3.3.1). Even human intentional actions, as well as goal-directed behaviours of other animals addressed at their own survival do not support an arbitrary generalization of finite final causes.

4.1.4 *Three Viewpoints on Created Reality*

The doctrine of Creation expressed in the *Summa Theologiae* that we have considered so far presupposes two further developments: (a) the treatment of God in His existence, His mode of being, and His operations[64], as well as (b) all the questions concerning His Trinitarian nature[65]. These two blocks are followed by (c) a specific part devoted to the procession of creatures[66], which we have addressed so far, and which is triadically structured as well. In particular, in this part (c), St Thomas deals with the following issues:

- The common source of any being, that is, God as the universal cause of everything[67];
- The distinction among creatures[68];
- The conservation and government of creatures[69].

When we consider God as the universal cause of everything we imply that every finite being *shares* or participates in this fundamental reality that is *being*. We recall that St Thomas conceives of God as the only being in which His existence coincides with His essence[70], and therefore as the only *being par excellence*[71] (Sec. 3.4). Thus, this part of St Thomas' theology of Creation deals with what everything has *in common*, which can also be considered as the ground of the intelligibility of everything, especially taking into account that the whole of Creation is regarded as intelligible (Subsec. 4.1.1). This is also the ground for the theological methodology

---

[63] See on this point Auletta 2011a, Chs. 8-11.

[64] Aquinas *S. Th.*, I, qq. 1-26.

[65] Aquinas *S. Th.*, I, qq. 27-43.

[66] Aquinas *S. Th.*, I, qq. 44-119.

[67] Aquinas *S. Th.*, I, qq. 44-46.

[68] Aquinas *S. Th.*, I, qq. 47-102.

[69] Aquinas *S. Th.*, I, qq. 103-119.

[70] Aquinas *S. Th.*, I, q. 3, a. 3.

[71] Aquinas *S. Th.*, I, q. 3, a. 4; *De Pot.*, q. 2 a. 1.

of the *analogia entis*[72]. The second point above rather deals with the amazing diversification of things in nature. We recall here that, according to St Thomas, what is one and simple in God is multiple and divided in finite beings[73], which also implies that the universe would not be so perfect without such a differentiation[74]. The third point above regards the fundamental dynamism of the finite being. Indeed, starting from *quaestio* 105 that is about the changes induced by God in the creatures, the rest is entirely devoted to the multiple effects among creatures, and in particular *quaestio* 106 deals with how creatures move or induce changes in other creatures in general, *quaestiones* 107-114 concern the operations of angels and demons, while *quaestio* 115 is focussed on the effects that bodily creatures have on each other (that is, their mutual interactions).

4.2 Reflections on the Three Principles

The previous examination shows some fundamental tenets of a traditional theology of Creation that should be taken into account when developing a theology of nature (Sec. 2.3). Here, we would like to summarize and extend some of the discovered possible connections between these theological tenets and the introduced principles of philosophy of nature, as these connections might become relevant for dealing with the 2nd, 4th, 5th and 6th problems[75] listed in Subsec. 2.3.7: the nature of contingency and its relations with necessity, the features of the final and temporal being, the relation between the creatures and the Creator seen from the viewpoint of the created world, as well as the monism-dualism issue. Besides the epistemological requirements that are necessary to avoid any kind of confusion among methodologies and competences[76], we emphasize that the combined effort to address these problems should be grounded on the assumption of a *coherence theory of truth* that has characterized Catholicism and a large part of the Christian

---

[72] As stated by Bochenski (1956, § 28), St Thomas' analogy is a similarity relation between two relations (an isomorphism) and as such does not require the knowledge of the relatives occurring in the other relation (which is a necessary requirement when the other extreme of an analogy is God).

[73] Aquinas *S. Th.*, I, q. 47 a. 1.

[74] Aquinas *S. Th.*, I, q. 47 a. 2. It is also interesting to note that here St Thomas hints at a certain *gradation* of created things.

[75] John Paul II 1988.

[76] Maritain 1934, pp. 93-95.

tradition (Subsec. 2.3.2). This does not imply the construction of an all-encompassing system of thought, but is rather the requirement for any Christian teaching to be consistent with all aspects of the reality of the world and of human life, as far as they are understood according to the state of the art. For example, the claim that God is the creator of the cosmos must be brought into consonance with our scientific knowledge of the universe[77] (Subsec. 2.3.1). Under this respect, it is worth noting that Thomas Aquinas in his *Summa Theologiae* did not argue directly from the principles of Revelation (the articles of faith) but constructed his argumentation (best described as *congruence*) by showing that some piece of knowledge about the world (as often examined in the philosophical-scientific[78] tradition) is consonant with, and even leads to, those principles[79]. We also stress that such a congruence is requested neither to science, which is always connected with a specific field of investigation, nor to philosophy, which, being an open critical enterprise, rather has the duty to show *incoherence* in theories and traditions and to deal with the formulation of cognitive strategies and methodologies that are finally addressed at solving whole but specific classes of problems (Subsec. 2.2.7 and Sec. 3.4).

The mentioned partition of the *Summa* (Subsec. 4.1.4) suggests a certain conceptual articulation that could be mapped to our three philosophical principles. We think that this is possible by establishing a correspondence between St Thomas' conception of God as the common source of every being and our principle of intelligibility, since, as mentioned, the procession of the finite beings from God is intelligible (Subsec. 4.1.1). The fact that all natural entities rest upon an ontological substratum that accounts for their regularities, provides an ultimate rational foundation for the philosophical and scientific enterprise[80] (Subsec. 3.2.6). As to this, Torrance[81], by supporting the idea that the ultimate ground of order is God's mind, also stresses that theology and science share a common presupposition.

We would also like to establish a further correspondence between St Thomas' conception of the distinction among creatures and our principle of the ontological

---

[77] Pannenberg 2008, p. 7.

[78] Aristotle as well as some Islamic authors were considered by St Thomas not only as philosophers but also as scientific authorities.

[79] Pannenberg 2008, p. 3. McGrath 2001-2003, v. 3, p. 141.

[80] Tanzella-Nitti 1997, pp. 17-18 and 26; 2004, p. 16.

[81] Torrance 1985b.

irreducibility of happenings, since this irreducibility is the source of variety in our world (Subsec. 3.1.8). In Whitehead's words[82], creativity is the principle of novelty. This brings us once again to the importance of contingency in theology, which is especially relevant for a theological reflection on biological evolution (Subsecs. 2.3.3 and 3.3.6).

Although there is a correspondence, there is also an inversion of our first and second principles with respect to the order followed by St Thomas' exposition in the *Summa Theologiae*. This is unlikely to be by chance since, following a systematic theological order with God as the primary Cause, the *Summa* deals first with what creatures share, i.e. their dependence on God, and then with their differences[83]. On the contrary, we have ordered our principles by starting from empirical evidence and therefore from the experience of happenings in nature, and, in this perspective, the distinction among things is epistemologically prior to their commonality and regularity.

There is also a certain correspondence between St Thomas' understanding of the conservation and government of creatures (Subsec. 4.1.3) and our third principle about the dynamical interplay between happenings and regularities. Here, we would like to add some words by considering the relation between form and matter as well as between contingency and necessity in secondary causation (Subsec. 4.1.2). In this respect, according to St Thomas[84], we should not look for the cause of corporeal forms in any immaterial archetype (like God's mind conceived as ensemble of many ideas in the sense in which Erigena understood it), but in something that is already composed of matter and form (Subsec. 4.1.1). The intelligible forms are created in an indissoluble unity together with material substances so that the latter are caused, not as emanations from some pre-existing immaterial archetype, but through the dynamical process bringing informed matter from potentiality into act; thus, the realized form cannot be something that is already presupposed (Subsec. 3.2.4-3.2.5). This is in agreement with our previous analysis of necessity (Subsec. 3.3.1) and with the ubiquitous fact in nature that even formal constraints, being potentialities, are to be activated and therefore fully realised through dynamical processes (Subsec. 3.3.3).

---

[82] Whitehead 1929, p. 21.

[83] In the words of Whitehead (1929, p. 348) we can say that the world is characterized by mutual exclusivity among finite things, while God is all-embracing.

[84] Aquinas *S. Th.*, I, q. 65, a. 4.

Summarizing, such a framework suggests an original explanation of creation's openness to the multiplicity and richness of phenomena as well as the capability of natural beings to give rise to an almost infinite number of connections and different relations with other natural entities (*in natura est alterum propter alterum*), thus grounding the process of becoming of the natural world[85]. Moreover, the issue of the gradation of finite beings supports the idea of the irreducibility of higher to lower forms of beings, which is the content of the philosophical perspective of *emergent monism* (Subsec. 3.3.5). Strong emergent phenomena such as life should be interpreted as resulting from the occurrence of natural secondary causes, being at the same time willed and supported at the level of primary Causation. It is therefore reasonable to guess that the Creator longs for an emerging universe with a real dimension of spontaneity, accident and large temporal duration[86].

4.3 Reflections on Intelligible Emanation, Design and Mechanism

We have considered above (Subsec. 3.3.6) the impossibility of conceiving God (and His providential care of the universe) in terms of a design (see also Subsecs. 2.3.3-2.3.4). Properly speaking, we have a design when people, especially engineers, try to build some prototypes of certain objects. This is evident, for instance, when studies are developed in the first step of industrial production. Since no finite being has the power of creation, but at most has the intellectual ability to arrange different things so that they work according to our goals, it is obvious that we need to consider some details of the final product in those prototypes; obviously, not all the details and not in all their specifics. However, to consider some details is somehow fundamental for providing a model that is also helpful for further stages of the object production. It is also clear that when supporters of the intelligent design speak of organisms as designed, they are again referring to some specific details, like having a purposeful heart for pumping blood, having a very complex brain, and so on. The application of this kind of representation to God's ideas is one of the main misunderstandings in both philosophy and theology. As we have said, God's ideas cannot have any specification that would anchor them

---

[85] Tanzella-Nitti 1997, pp. 6-9.
[86] Haught 2006. p. 71. Coyne 2007. We note that also the theologian D. Edwards (2010, pp. 2-8) speaks of some fundamental characteristics of our universe that bear resemblance to our three principles. Namely, he speaks of (i) evolution (dynamicity), (ii) relationship and patterns, and (iii) integrity of any pohenomenon, which can be understood in terms of our third principle.

to certain solutions and even to a certain universe (Subsec. 4.1.1 and Sec. 4.2). This misunderstanding can even be found in the father of the formal world, namely Plato, who famously asked in his dialogue *Parmenides* whether there are ideas of Socrates' hair, or of mud[87].

The problem is ill-posed. Even if the world were totally mechanistic, this would not diminish God's power by a single degree. Since God's power and modality of causation do not interfere at all with any finite mode of causation (Subsec. 4.1.2), to assume that a mechanist world would represent a diminution of God's power shows how misleading the intelligent-design viewpoint is, which is still upheld by many theologians today and risks transforming God into an anthropomorphic agent[88]. On the other hand, very skilled engineers are able not only to build mechanical devices but actually also to copy biological molecules and organs like it is currently the case for the production of artificial DNA or hearts. It is even envisageable that humans could build new complex systems in analogy with biological organisms. So, there is nothing mysterious or supernatural in the accordance that there can be between several parts of a system or an organism. What is instead truly extraordinary is the fact that such a very fine-tuned harmony of parts and organs comes about *spontaneously* in nature arising from a dynamical process grounded in intelligible emanation of the entire universe from its first Principle. Creatures show dynamical potentiality not only to realise their full expression (as it is the case for organisms reaching their mature form in a developmental process) but also to induce changes and to activate formal constraints that result in new and unexpected forms of being (as it is common in the evolution of biological species, Subsec. 3.3.1).

If so, we would strongly suggest avoiding any attribution of sense or meaning to any detail or part or even temporal segment of our world. To do so would be contrary not only to the spirit of scientific enquiry but also to a correctly understood theology whose primary concern is to avoid mixing up God's action with any form of finite causation, included finite formal and final causes (Subsec. 4.1.3). This is why, in the context of the theory of evolution, we strongly support Haught's words: "If evolution has a meaning, it would be embedded in the narrative depths of life rather than in isolated instances of complex design that float occasionally to the surface. It is to the underlying story, not to isolated complex systems or elaborately structured molecular states, that we should look to make sense of evolution. But it is the nature of drama that it takes time to unfold, so that whatever meaning it

---

[87] Plato *Parmenides*, 130c-d.
[88] See Edwards 2010, pp. 45-47.

carries it cannot unveil itself fully to our intellectual curiosity or scientific observation at any present moment. So, if life is essentially a drama rather than a factory of designs, it is arrogant and short-sighted for anyone simply to declare it meaningless because of the imperfection of present and past evolutionary adaptations"[89].

Coming back to strict theological considerations, Creation is a multifaceted relation that involves many aspects. The first one is the power to bring something into existence (Subsecs. 4.1.2-4.1.3). The second one is the capacity to conceive something that could therefore also instantiate objective intelligibility (Subsecs. 3.2.6 and 4.1.1). However, the issue of Creation is also concerned with its goodness: God not only creates something, but also wills something. This goodness is expressed by the universe's "responsitivity", i.e. its participatory character (Subsec. 3.3.4) that makes it not only an object but something having in itself the seeds of further action and intelligence; something, said in a word, that could be able to respond to a call and also, at another level of complexity, to show moral responsability. Finally, creation shows an intrinsic beauty due to its balance between order and disorder (Subsec. 3.3.6) and indeed it has been appraised by God. This complex relation between Creator and creation has been summarized in St Thomas' doctrine of "*transcendentals*", according to which *ens, unum, verum,* and *bonum* are perfectly equivalent in God but still preserve a relation across the differentiated intelligible emanation that we call Creation[90], which implies the impossibility to endow any finite being or aspect of the world with a sort of supernatural meaning. It is only with this proviso that we can consider, with St Thomas, our universe to be the best of all possible worlds[91]. Thus, by reflecting on the created nature, on its order and beauty, we can infer and admire God's wisdom and goodness although in an indirect way[92].

4.4 Reflections on Causality

In the previous sections, we have seen how fertile St Thomas' theological doctrine is, especially concerning the concept of intelligible emanation and the distinction between primary Cause and secondary causes. Although we have also

---

[89] Haught 2010, p. 59. These considerations and what said in Subsecs. 2.3.3-2.3.4 would suggest a more balanced evaluation of Teilhard de Chardin's contribution than that provided in Ch. 11 of Haught's book, although we agree that is a merit of Teilhard to have shown that thought and spirituality can emerge as a result of an evolutionary process.

[90] Aquinas *De Ver.*, q. 1, a. 1. On God's Unity see also Maimonides *GE*, I, § 57.

[91] Aquinas *S.Th*, I, q. 25 a. 6

[92] Aquinas *Cont. Gent.*, b. II, ch. 2. See McGrath 2001-2003, v. 1, p. 172.

considered this problem as being connected with the fundamental dynamism of creatures, we wish to develop this order of questions in relation to two important problems: the doctrine of the *Creatio continua*, where Creation is a modality of action proper to God and only to Him[93], and the notion of *actus essendi* grounding the existence of any individual finite being. Indeed, secondary causes far from impairing God's continual Creation are an essential part of it, and their activities are not to be conceived at the same level as that of the Creator[94]: the primary Cause causes *per modum creationis*, the secondary causes cause *per modum informationis*.

Let us point out that secondary causes act according to what they have received by the primary Cause, and by virtue of that they are effective both in the order of becoming and in the order of *esse*[95]. Relative to the latter point, as stressed by St Thomas, secondary causes can be causes of the being of other creatures only to the extent to which they can help to *draw out* (*educere*) the form from the matter, that is, in activating potentialities[96] (Subsec. 3.2.5 and Sec. 4.3). If we confine our examination to efficient causation only, this doctrine becomes fully incomprehensible. However, we have stressed (Subsecs. 3.3.1-3.3.3) that the fundamental dynamism ruling all natural processes determines the activation of formal constraints at all scales, which are in turn able to back-react on further processes and systems. Nonetheless, by virtue of the correlations existing in nature, such an immediate effect spreads over the activation of formal constraints that in turn participate in the emergence of new forms of reality and thus to the constitution not only of individuals but also of *new kinds* of individuals (as it happens with speciation thanks to teleonomic circuits). According to the view presented here, it would be obsolete to consider St Thomas' ontology as being anchored to a fixist vision of the world[97], precisely because the government of creatures by God is performed *through* the nature of any and all the created beings concurring to multiple effects in the generation of new kinds of reality[98]. This also means that the reality of every finite being depends on a huge amount of relations and interactions that it has with other finite beings. This would be barely understandable from an Aristotelian pespective that makes relations accidental. However, we have

---

[93] Aquinas *Cont. Gent.*, b. II, ch. 17, 20.

[94] On this point see also Pannenberg 2008, p. 93.

[95] Aquinas *Sup. Sent.* II, d. 1 q. 1 a. 4.

[96] See also Selvaggi 1964, pp. 61-62.

[97] Salucci 2011, §§ 6.3-4.

[98] Tanzella-Nitti 2004. See also Pannenberg 2008, p. 39.

seen (Subsec. 3.2.3) that St Augustine understood relations as having an impotant ontological status. The trinitary God consists of unity *and* relations. We can then understand that relations are also fundamental for finite beings, although in an analogical way. Indeed, relations among the Trinitarian Persons display the autonomy of God, although showing the dynamicity of the spiritual life. The relations characterizing finite beings show their contingency and heteronomy, but also their fundametal openness.

Obviously, it is impossible for finite beings to act directly on formal constraints, and this explains how justified is the common scientific assumption that (finite) causes act immediately only on observable properties or states of certain systems, so that scientific research is not concerned with their existence as such that involves those correlations[99]. This is one of the reasons why philosophers like ancient atomists[100], but also modern corpuscularists like Locke[101], upheld that corpuscles constituting matter are permanent (that is, neither produced nor annihilated during the history of our universe) and only their properties, or states like motion, could change due to the interaction with other beings. Traditional atomistic and corpuscular theories tried to reduce every difference in our universe to a few number of elementary constituents, whose combination would account for everything else (the so-called secondary qualities as well as biological functions or the human mind: see Subsec. 2.2.5). Obviously, this was a radical choice for avoiding to deal with any kind of interdependencies.

When we speak of existence, the biggest difficulty (which is a manifestation of the universe's variety) is the irreducibility of *individuals* and therefore also of their *existence*[102]. Indeed, it is well known how difficult is to account for the popping up of an individual in our universe[103], and this explains why many ancient and modern philosophers were led to adopt such an economic solution and considered individ-

---

[99] See also Coyne 2007, p. 14.

[100] See Cudworth's (1678, I, Ch. 1, § XXVIII) examination on this point. Cudworth's work is an interesting reaction to the new mechanist metaphysics. The interest lies in the fact that he does not deny the existence and relevance of mechanical causes but nevertheless stresses the full compatibility of this view with the assumption of the existence of non-mechanical principles and factors that he, following Plato, certainly interprets in terms of separated substances (1678, I, Ch. 1, § XLI) but being simultaneously aware that ancient Platonism dismissed mechanical explanations too easily (1678, I, Ch. 1, §§ XLIII-XLIV).

[101] Locke 1689, IV, Ch. 3, § 16.

[102] Leibniz 1686b, p. 433; 1689b, pp. 519-520; Auletta 2011a, Chs. 10-11.

[103] Hamilton 1859, I, Lect. IV, pp. 66-67.

uality simply as an illusion or an epiphenomenon derived from a momentary aggregation of corpuscules or atoms[104]. We should however distinguish between the origin of individuals and the role that the acknowledgement of their existence plays in providing sufficiently sophisiticated explanations of nature. Indeed, we are able to understand more and more the dynamic interplay between organized entities or wholes and their parts (as shown by our third principle of philosophy of nature). To take an example, it is difficult to explain the complex social interactions occurring among humans (especially the back effects of social relations on individuals and, reciprocally, of individuals on groups) without taking into account the existence and function of individual entities. As we have mentioned, the problem of the irreducibility of individuals was well known to Aristotle, who introduced the concept of τόδε τι[105], which was also developed in the Middle Age tradition, for instance with the concept of *haecceitas* (Subsecs. 3.1.8 and 3.3.7). Such an issue leads us back to the irreducibility of our experience of the world to the cognitive structure of the human subject (Secs. 2.1 and 3.1). In other words, science is today obliged to acknowledge certain realities (a physical structure, a living organism deploying certain functions, a person, or a community) as a *given* from which to start the investigation, being unable to explain in a complete way how such a reality arises at all. This problem is also true for philosophy, which is forced to acknowledge the existence of individuality as an ultimate fact and to introduce a heuristic principle to explain the *general* fact that individual variety and random happenings exist (see also Sec. 3.4). On the other hand, by virtue of the distinction between primary Cause bringing individuals to *existence* and secondary causes determining *specific effects*, included the activation of coustraints in other beings (Subsec. 4.1.2), theology may help us to find full clarity on this point. Thus, we may understand that, although individuality is irreducible and grounded on God's continual Creation (*actus essendi*), the existence of every being also depends on the way in which interactions and relations that it entertains with many other be-

---

[104] Spinoza says: "Cum corpora aliquot eiusdem, aut diversae magnitudinis a reliquis ita coercentur, ut invicem incumbant, vel si eodem, aut diversis celeritatis gradibus moventur, ut motus suos invicem certa quidam ratione communicent, illa corpora invicem unita dicemus, et omnia simul unum corpus, sive Individuum componere, quod a reliquis per hanc corporum unionem distinguitur." (1677, pp. 99-100). This is also the reason why most philosophers mix ontological reference with inference (Subsec. 2.2.3) although we have remarked that we can also reach ontological conclusions starting from inferences (Subsec. 2.1.1). However, the assent to these conclusions is not itself a matter of inference, as clearly stated by J. H. Newman (1870, Chs. 3-4).

[105] Aristotle *Cat.* 3b10-23.

ings and ultimately (although obscurely) with the whole universe suddenly coalesce in a single systemic unity, which, theologically speaking, is the act of being. Indeed, only God's existence is absolute, whilst every finite being's existence depends on specific contexts and conditions, precisely those ascertained by empirical sciences.

4.5 Reflections on God's Action and Humanity

Creation should be understood as a real relation of all and any created reality with the Creator and as His continual sustaining of the existence of finite beings (Subsec. 4.1.2 and Sec. 4.4). The doctrine of continual Creation implies that such creative relation concerns not only the "beginning", but the whole course of creation's history. The *particular* and *specific* ways in which God continuously sustains His creation conferring existence to it are not easy to grasp or to express[106] in terms that completely avoid the risk of anthropomorphizing God's action or to imagine Him intervening directly in the universe as a being among other beings (Subsecs. 2.3.3-2.3.4 and Sec. 4.3). In other words, God's providential plans remain obscure in their details[107].

It is worth mentioning here the attempt developed in the last decades by R. J. Russell with his doctrine of a "non interventionist objective divine action"[108]. One of the main problems the author addresses is how to reconcile the issue of divine action with the autonomy of the universe and its fundamental laws and characters. The author finds in the ontological indeterminism suggested by quantum mechanics a way to understand God's action towards creation in a non-interventionist and generalized way, thus upholding a theistic perspective and avoiding the danger of deism. Although in the previous examination we have faced the problems on the basis of St Thomas distinction between primary Cause and secondary causes, we strongly share the worries and exigencies beneath the author's proposal.

Another relevant issue when speaking of God's action, is that of miracles. Far from claiming to exhaust such a difficult matter, we emphasize their testifying dimension and meaning[109] in order to avoid conceiving them as *mere* portentous happenings representing the overwhelming power of a deity again possibly conceived in mythological terms[110]. For example, Maldamé invites us to consider

---

[106] Tanzella-Nitti 2009, p. 45.

[107] With regards to God's special action (or special Providence) see Saunders 2002.

[108] Russell 2008, Ch. 4-6. See also the enlightening PhD dissertation Silva 2010.

[109] Cf. Lk. 4, 9-12.

[110] Tanzella-Nitti 2009, pp. 179-182.

miracles not as a rough intervention of God in nature but rather as having the function to *communicate* God's goal of salvation, to reveal God Himself, to confirm His message, and to be beneficial[111]. We recall that Christ's miracles are always due to the faith of the receiver and cannot be understood as showing God's power on nature[112].

In this sense, the issue of miracles may provide a bridge between God's activity as the continuous Creator of our world and the fundamental dimension of His action towards humanity[113], namely, His Revelation through His Word and Christ's life as the incarnated Word (an exigency also endorsed by Barth, see Point 3 of the introduction to Sec. 4.1). This particular Revelation and Providence addressed to the being *capax Dei* is not in contradiction but in continuity with the other form of God's self-revelation, that is, Creation. Therefore, if the particular modalities of His continual assistance to creation are, as such, very hard to grasp, there is a fundamental place in which we may well recognize God's activity in positive terms: the ways in which He reveals Himself to us as the Primary source of all being, truth, goodness and beauty (Sec. 4.3). We would like to emphasize that[114]: (1) the soteriological message is obviously addressed *only to us* as human beings living in a condition determined by sin; (2) salvation is addressed to us as persons *by a personal God* and is ultimately allowed by the redemption fully accomplished in Christ.

The first consideration implies that for an integrated cognitive enterprise like the one we wish to pursue, the specific symbolic activity of humankind can be regarded as the natural character that renders us *open to the supernatural*, and eventually capable of answering to God's Revelation nothwithstanding the intrinsic limitations that are derived from our condition, thus establishing an actual communication. This brings us to the second consideration, which underlines how such a communication starts as well as what it means. It is well known that Einstein refuted to admit the existence of a personal God[115], but he deeply acknowledged the super-personal dimension of the deity. As we have mentioned (Subsec. 4.1.3),

---

[111] Maldamé 2003, pp. 189-90. Tanzella-Nitti 2009, p. 174. In this way, miracles announce God's will to save everybody and the comping of His Reign (Edwards 2010, pp, 18-20).
[112] Tanzella-Nitti 2009, pp. 127-128.
[113] Tanzella-Nitti 2009, pp. 194-195
[114] Tanzella-Nitti 2009, pp. 31-44.
[115] See Jammer 1999, pp. 47-51 and 94-97.

the Christian theologian Paul Tillich carefully examined Einstein's position[116]. Among other things, Tillich pointed out that if God needs to be "super-personal", as Einstein acknowledged, He cannot be an impersonal "It" (Being) only, but must also be a "He" (a person) and even exceed these qualifications (Subsec. 2.3.7 and Sec. 3.4). If it were not so, this would transform God into a sub-personal entity[117]. Now, since we experience that in our universe there is emergence of life and mind, we cannot avoid attributing life and mind also to God, although in an analogical way. We shall consider in the next chapter how fruitful these ideas are for a renewed anthropology. We shall start, however, by stressing what they imply from the point of view of our relation with God.

Being a person in a sublime sense, God is also an agent in an eminent sense and it is quite natural that it is God that reveals Himself to every one of us: He makes the "first move" towards us. We may actively search for God everywhere; this is an important part of our humanity. There is an ever-present need and hope to move from obscurity to clarity about reality and its meaning and value, and this is the ultimate horizon in which knowledge and redemption coincide[118]. Finally, we find Him in a disinterested act of free-giving that we are willing to accept. This is a real *action* undertaken by God, which in itself does not change our natural bodies or the physical world surrounding us but may definitively change our lives and indirectly also the world, by means of the actions we, in our turn, as secondary causes provided with rationality, can undertake according to the answer we give to God's call. Clearly, the spiritual dimension of salvation also projects us towards an order of considerations related to what is beyond our conception of, and even our persistence in, the natural dimension in which we live. Therefore, both God's action as the Creator and as the personal God who reveals Himself are complementary issues in a deep sense. Here, it is sufficient to recall the Gospel[119],

---

[116] Quoted in Jammer 1999, pp. 107-113.

[117] It is likely that the understanding of God as an impersonal It that also characterizes deism and can even be found in Draper (1875, pp. 121-140) reflects a stage of science essentially dominated by classical physical sciences in which "force" is considered as "indestructible and eternal".

[118] Haughey 2009, p. 45. As it has been recalled (McGrath 2001-2003, v. 1, p. 162), one of the first authors to argue in favour of the strict interrelation between redemption and Creation within the general framework of salvation was Irenaeus of Lyons, in response to the Gnostic conceptions about God's creative action.

[119] Jn 1, 1-3.

which teaches that through Christ everything has been made: indeed, the *Logos* through whom everything that is created took its form is the same *Logos* who became incarnated[120] (Subsec. 4.1.1). The centrality of Christ may be seen as the highest point of convergence of the discourses on Creation and Salvation. This would lead us to Christology as the very core of a theological reflection concerning the created nature (and the issue of evil and sin), a matter obviously falling outside the limited scope of the present work.

---

[120] Aquinas *Sup. Sent.* II, d. 1, q. 1, a. 6. As McGrath remarkably summarizes: "The ordering and rationality embedded in nature, and capable of being discerned by the human mind as created *in imagine Dei*, is embodied in Christ" (McGrath 2001-2003, v.2, p. 313; see also v. 1, pp. 24-25, 59-60, 98). See also Tanzella-Nitti 2009, Ch 5.

5

TOWARDS A NEW ANTHROPOLOGY

In Ch. 3 we formulated three general principles of philosophy of nature. In Ch. 4 we sketched a theology of nature by framing it in St Thomas' theology of Creation. In particular, in Sec. 4.5 we developed some considerations that addressed the question of humanity. Now we would like to approach this problem more extensively. It is well known that each of the three fields involved here, namely theology, philosophy and science, has abandoned certain topics in favour of the other two. For instance, the enquiry about the natural world is mostly confined to scientific developments, and any question about the possible meaning of our universe and the reason for its existence is confined to theology, whilst philosophy is accustomed to deal with epistemological, methodological and even metaphysical issues. The case of anthropology is totally different. Whereas we can say that the natural world is the proper (material) object of empirical sciences (Sec. 3.4), this is hardly the case for man. Here, any of the three fields can legitimately claim mankind to be its subject matter[1]. What is extraordinary, however, is that each of the three fields has developed its own approach mostly without taking into any consideration the contributions coming from the other ones. We shall first deal with the ways in which humanity is considered in those disciplines and then try to formulate an original approach. In this way we shall be able to deal not only philosophically but also theologically with the monism/dualism issue (see question 7 in Subsec. 2.2.7 and question 6 in Subsec. 2.3.7).

---

[1] As a matter of fact, many consider man and especially the mind as the proper object of philosophy (see Hamilton 1859, I, Lect. III, pp. 61-63). We agree with this statement, provided that we understand it in an inclusive way, that is, philosophy deals essentially with man and the mind but both natural sciences and theology can and should deal with these subjects in the ways that they consider to be most appropriate. We hope to clarify this point in the following.

5.1 Theology, Philosophy, and Science Dealing with Anthropology

The issue of human subjectivity treated by theological as well as philosophical anthropology involves a number of considerations that go beyond the extent of this work. Often, theologians feel that the anthropological dimension is not fully accounted for in the scientific framework. Let us consider here an authoritative theologian who is not hostile to science and even sympathetic with some scientific developments, namely J. F. Haught[2], who expresses a significant position with respect to many of the issues that we are interested in. In particular, we find that he raises an important problem, that is, the issue of the mediation or connection between anthropology and naturalism. In dealing with anthropology, we consider this connection to be the crucial problem at hand, since our culture is still dominated by a fracture stemming from the origin of the Modern Ages between scientific disciplines and humanities[3] (but see also Subsec. 2.2.3).

Although Haught attributes a positive value to the autonomy of the scientific enterprise, he also maintains that the knowledge acquired through natural sciences implies the conceptual decomposition of natural processes and is therefore focussed on atomic details, and thus rendered intrinsically unable to grasp the overall meaning of the universe[4]. Now, we certainly agree that science in itself cannot grasp such an overall meaning (Subsec. 2.1.6). We also understand that higher levels of organization and complexity are not "visible" in the treatment of many particular scientific problems. For instance, it is clear that when we are dealing with the synaptic connection between neurons, we overlook the overall emerging patterns of the brain, not to speak of the issue of consciousness. However, we do not wish to support the idea of a science consisting in a mere decomposition of phenomena, a view bound to a mechanist research strategy that, although still dominant today (Sec. 1.3), is likely to be overcome in the future. There is perhaps a terminological difference here in that Haught uses the term *naturalism* as a synonym for "scientism" or "reductionist science". Indeed, Haught explicitly refers to an understanding of naturalism deeply linked to materialism and to an ontological reductionism[5]. Obviously, we agree on this and we also explicitly discarded this conception; however, we also showed the possibility of a new understanding of naturalism (Subsec.

---

[2] Haught 2006.
[3] Snow 1960.
[4] Haught 2006, p. 123.
[5] Haught 2006, pp. 84-85 and 118.

2.2.5) and how to justify our view on naturalism through a philosophical reflection appropriately taking into account some of the current and most urgent scientific concerns, especially with regards to emergent and complex realities (Ch. 3).

Nevertheless, Haught also feels the urge to widen the scope of investigation when dealing with mankind. Indeed, he argues that an enlarged and enriched empiricism should directly involve cognitive acts characterizing humans (like to be attentive to some content or meaning, to understand such a meaning, to be critical and so to judge about it, or even to make moral decisions) as well as the phenomenological and subjective experience (as expressed, for instance, in affectivity, intersubjectivity, narrativity, and aesthetics) in order to account for our mental dimension, integrated personality, and desire of truth and meaning, which make a wider intelligibility of the world and a theological worldview possible[6]. Although the author maintains that these cognitive acts and phenomenological experiences of the subject do not imply any kind of dualism[7] (Sec. 2.1.6), he also stresses that they cannot be dealt with in a traditional scientific framework. We remark here that on the basis of the general conceptual framework we are proposing, we should avoid a generalized methodology centred on introspection and take into account empirical facts and results that come from the most recent scientific investigation (Sec. 3.4): natural sciences, especially when focussing on the relationships between consciousness and the complexity of the brain, may play a significant role in tackling issues like intentionality, purposeful behaviour, and free will as relevant aspects of subjectivity, thus also contributing to the "wider and deeper empiricism" invoked by Haught himself and so providing a fruitful bridge to the issues raised by him[8]. As a matter of fact, many scholars deeply concerned with cognitive neurosciences are making remarkable theoretical and philosophical efforts in explaining the natural and cognitive roots (or at least the context) of issues like those emphasized by Haught, within a scientific-based enquiry about human-specific symbolic activity and without devaluating the relevance of these issues in tracing them back to physical-chemical elements and interactions[9]. Symbols have this extraordinary character that makes them different relative to any kind of natural signs (which are used ubiquitously in any domain

---

[6] Haught 2006, pp. 32-54 and 96.

[7] Haught 2006, pp. 51, 86, 88, 93, 96.

[8] Haught 2006, pp. 124-126 and 128.

[9] Damasio 1999; Jeannerod 2006; Changeux 2009. See Auletta 2011a for the general characters of symbolic activity.

of life): they do not represent the contents they refer to[10]. For instance, the word "tree" does not provide any kind of representation of a tree and for this reason cannot be understood by somebody who does not already know English. This is also true for mathematical symbols. This character is what allows us to speak of things of which we cannot have a perceptual representation, like black holes, four-dimensional space, virtue, philosophy, infinity, and God. However, this is also the reason why symbols, when used for referring to things that can be experienced, need the biological substrate of the brain for producing an adequate representation. Therefore, we cannot separate those mental acts and phenomenological experiences at the center of Haught's work from such a substrate.

Moreover, to philosophically deal with anthropological problems without the contribution of science and theology one is a great mistake. Once the scientific enquiry was considered to be irrelevant for dealing properly with human fundamental questions, and theological perspectives were left apart in the name of a so-called secularized cultural enterprise, philosophy ended up in delineating a fragmented subjectivity inserted in constantly changing social constructs[11], with the consequence of a diffused philosophical relativism[12] (Subsec. 2.2.1). Unfortunately, in present times we often find philosophy as practiced in a sophistic manner in which, for instance, theses and arguments are assessed according to the meanings that the words composing single statements may or may not assume in a number of theoretically possible contexts, which is likely an abuse of the so-called *linguistic turn*, an approach ultimately stemming from Wittgenstein's famous *Philosophical Investigations*[13].

In sum, being interested in a significant exchange with scientific disciplines, we think that it is necessary to consider the unity of the human condition[14] in its physical, biological and psychological components without confining the issue of humanity within the limits of a specific investigation field, although many particular empirical results can be very helpful in such an enquiry[15].

---

[10] This is why Leibniz (1684) thought that symbolic thinking was blind. See also Hamilton 1859, III, Lect. X, pp. 179-183.

[11] Foucault 1969. Lyotard 1992.

[12] *Caritas in Veritate,* § 31.

[13] Wittgenstein 1953.

[14] Ladaria 1995, pp. 139-140.

[15] Auletta 2009b.

The physical component cannot be overlooked at the very least because it concerns our body directly and the undeniable fact that the human being is continually exposed to a physical context since its birth. Notably, the interactions we entertain with the environment are of the highest relevance in the formation of our experience of the world from which our cognition and knowledge stem[16] (Subsec. 3.1.6), constituting a source of stimuli and data for higher acts of interpretation. Therefore, even if it is basically related to the material constitution of the human being, the physical component is a field of interest also for disciplines like psychology and cognitive sciences.

Moreover, the typical character of humanity, namely the ability to establish culture and political societies, needs to be considered on the background of biological phenomena. We have already stressed that organisms are able to mould the environment according to their needs[17] (Subsec. 3.3.1). Animals have developed the highest forms for doing this and humanity emerged from such background. We are not speaking here of the specific forms that such an animal condition may take, but of some necessary requirements for the emergence of intelligence: we cannot have intelligence without the ability to represent the external world, but representations have in turn evolutionary emerged as a consequence of the necessity to monitor the effects produced on the external environment by the animal's motions. Indeed, the brain arises as an organ for controlling motion and consequently to produce representations[18]. Therefore, culture can be understood as a form of higher adaptation to the environment. While other animals try to transform their immediate environment, humans are even able to reinvent and model new environments as well as to intervene on the modalities of heritage: cultural transmission is indeed a new, non-genetic form of transmission to subsequent generations. Such a connection with the biological background is also what justifies our statement that religious traditions (which are a typical cultural phenomenon) although various, cannot be taken to be arbitrary, since they correspond to specific and successful forms of life (Subsec. 2.3.1). We do not need to recall how

---

[16] As shown in the path-breaking work of Piaget (1936 and 1945).

[17] Auletta 2011a, Ch. 8.

[18] Jeamerod 2006; Auletta 2011a, Ch. 12. This was perfectly clear to St Thomas, who stressed that also other animals need representations, and not only of sensible things that are immediately present but also of past events that they need to memorize and preserve (Aquinas, *S. Th.*,

unfruitful is a mechanist scientific research strategy that continues to consider an organism as exactly what it is not, that is, an artefact.

The psychological component of humanity represents an immense field of investigation. We need to consider here not only the mental sphere and the cognitive processes, but also the subjectivity involved in the phenomenological experiences underlined by Haught. Some of these issues have also been scientifically addressed with remarkable results, as in the case of the human affective dimension that is now increasingly acknowledged as relevant even for rational decisions and elaborations, and crucial for all social interactions that are at the root of culture[19]. What we would like to add here is that the psychological component as a whole (or in one of its sub-components) has very often been identified with the distinctive character of the human being. There are indeed many valid reasons to think so, yet we hope to have contributed to show that a unitary view should be achieved, and therefore that the problem is how to think and investigate the constitutive links between the psychological components and the other two. Even if the physical and the biological dimensions obviously belong to other natural beings, we think that in the human being they are expressed in a very distinctive way. We shall now address in particular the mental dimension from a philosophical point of view.

5.2 The Physical, the Formal, and the Mental: Philosophical Approaches

As mentioned at the beginning of the present work (Sec. 1.2), St Thomas' contribution may be considered as a fruitful compromise between the scientific research strategy grounded in a Platonic school of thought and the Aristotelian causality-based approach to nature (Subsec. 3.3.3). We saw that, in ancient times, science used to be essentially descriptive and based on Platonic ideas, or on what we can call the *formal world* (the esemble of mathematical, logical and idealized constructs), as an ultimate criterion of objectivity. Aristotle tried to take into consideration a radically different point of view, that is, to deal with the *natural world*, encompassing changes that are empirically ascertainable, and therefore offering the first attempt at pointing out explanatory mechanisms (causes) inherent to that world (Subsec. 3.3.7). Yet the times were not ripe, at least since empirical evidence was insufficient to build a consistent theory of dynamical processes, not to speak of a true research strategy. In such a context, it is important to clarify that the Aristotelian philosophy was not a sort of reply to Plato. As a matter of fact, Aris-

---

[19] Damasio 1999.

totle answered to a completely different question with respect to Plato: the latter asked about the way in which material substances are related to the source of their intelligibility (i.e. ideas), while the main problem of Aristotle was how material entities are subjected to formal constraints, and eventually final causes, in their dynamic changes and in the course of their mutual interactions. These two approaches are not necessarily incompatible but respond to the treatment of mathematical and universal ideas, on the one hand, and to the issue of the constitution of natural entities (especially organisms), on the other (Sec. 3.4). It is not by chance that in any treatment of biological systems, Aristotelian categories always come out, especially when epigenetic-developmental problems are involved[20] (Subsecs. 3.3.1 and 3.3.7), while any long-lasting philosophy of mathematics has been essentially Platonic[21].

St Thomas' understanding of formal causes as embodied in material substances, the doctrine of *potentiality*, the doctrine of the four causes, and a certain acknowledgement of the centrality of dynamics are all elements derived from Aristotelian philosophy, which also plays a relevant role in our own philosophical approach[22]. On the other hand, the particular understanding of God's mind and the universal ideas are obviously of Platonic derivation, and even play an important role for the principle of intelligibility. We may suggest that St Thomas' distinction between Primary and secondary causes allows for a compromise between Platonism and Aristotelism, based, from a theological perspective, on the notion of Creation (intelligible emanation) that was not available in the ancient Greek milieu (Secs. 4.1-4.4).

The theological discourse on intelligible emanation means that God's power and intelligence are strictly intertwined in the process of Creation. This implies that there is in fact a connection between Plato's ideas and Aristotle's formal causes and constraints[23]. Philosophically speaking we have found this connection in the concept of information (Subsec. 3.2.4) whilst, theologically speaking, we find this linkage in the interconnection between God's power and God's intelligence although we recall that there is no plurality in God's mind so that the specification of idea is only through such a process of Creation in which intelligible forms are

---

[20] Auletta 2011a, Ch 11.

[21] Heller 2005; Zycinski 2006c.

[22] Auletta 2006a; 2008; Auletta–Torcal 2011.

[23] The extraction of intelligible forms from the perception of the material object is a complex subject matter that was already present in Aristotle but much more in the Thomistic tradition (see Fabro 1941b, pp. 111-120 and 232-253). It is possible that Aristotle (*An. Post.*, 99b26-100b5) generated a confusion between two different issues: on the one hand the progress of

created in indissoluble union with bodily substances. Now, the human being can be seen as an intellectual being created by God (*imago Dei*)[24], so that the issues of the order of the created natural world and of the humans striving to the interpretation of the empirical experience clearly become the two distinct but inseparable sides of the coin when speaking of intelligibility. Understanding such a strict relation requires a reflection on the peculiar intellectual capability of the human subject, that is, in our view, an appropriate account of the *activity* of the mind, where a higher level of connection between the world of forms and nature may take place[25]. Our self-elevation to the understanding of the ultimate truths, which transforms us into that rational animal of which we are embryonic forms[26], is a sort of intellectual process running backwards relative to the ontological process of Creation. We try, so to speak, to guess what exactly the process of intellectual specification *could have been* that gave rise to our universe. To use the words of C. S. Peirce, intellectual life shows a tendency to value existence as the vehicle of forms[27]. We do this in a *discursive* way, whilst Creation itself is the expression of a unique power without any articulation. Therefore, the formal entities grasped by the human mind (that we try to identify as species or classes of things) are in a certain sense subjective, even if they are shared among humans. Therefore, they do not lack intersubjectivity. Moreover, they do not even lack a certain objectivity because they truly represent the way in which we, although in a discursive way, try to grasp the ultimate foundation of any intelligibility and so to approach the Primary principle of everything, what we may call the ultimate Truth (Subsec. 2.1.4). For this reason, this could be shared also by other intelligent beings in the universe. This view seems to answer to Einstein's worries about intelligibility. Although he believed in the intelligibility of our universe, since, on the other hand,

---

knowledge through a progressive generalization but also specification (which, as we shall explain, is hypothetical), on the other the specific form of inference called induction (see also Subsec. 3.1.2). The former process proceeds by combining principles and experience (see also Subsec. 2.1.2) and therefore makes usage of all the three basic forms of inference, i.e. natural deduction, abuction and induction, although perception is obviously crucial, constituting the basis of our experience (on these matters see Ferejohn 2009).

[24] Aquinas *Cont. Gent.*, b. II, ch. 46.

[25] See Tanzella-Nitti 2009, p. 42.

[26] Peirce 1898b, p. 50.

[27] Peirce 1905b, p. 347. Reciprocally, it is true that "a world of chance is simply our actual world viewed from the standpoint of an animal at the very vanishing-point of intelligence" (Peirce CP 6.406).

could not rationally justify this insight, was finally puzzled about the accordance of our theory with reality (Subsec. 3.2.6). They agree but cannot coincide. They would if we could see reality from the point of view of God, what we cannot. In the following sections we shall show that we can and must strive for overcoming this unbridgeable gap confident that we approach to truth (Subsec. 4.1.1). We have first tried to philosophically and theologically justify intelligibility *ex objecto* and now we have to approach the matter from the subjective standpoint.

Nowadays, especially when theological considerations are set aside, one could be inclined to attribute reality only to the physical and at most to the mental dimension[28], regarding the formal one as not having any kind of autonomy or reality. Of course, theoretical elaborations originally come from the creativity (mental activity) of those scholars who propose them and as such they have to be subsequently faced with empirical data, observations and experimental outcomes (i.e. with the physical dimension), which in turn test our elaborations along the historical development of the empirical sciences. However, it should not be forgotten that our theoretical elaborations are also constrained *independently* of their application to empirical reality, especially regarding their hypothetical character: mathematical and logical requirements must be satisfied in any sound elaboration. It is possible that there are other kinds of formal requirements since we see that, in very different fields, starting from very different viewpoints, perspectives and problems, scholars often reach similar theoretical solutions. This suggests that the "existence" of a world of formal entities should not be underestimated[29]. Consequently, we do not agree with the philosophical standpoint expressed by F. Suárez that became paradigmatic for the Modern Ages: according to him essences are merely ways to consider, in our knowledge, the things of the universe, which are the only ones that can be said to truly exist[30].

In the progress of our knowledge, we try to understand its objects more and more, which also means to understand them in more general and abstract ways[31]

---

[28] A side-effect of a one-sided cultivated natural science (Hamilton 1859, I, Lect. II, pp. 35-36).

[29] See also Peirce 1905b, pp. 354-57. This was considered by Peirce as constituting one of the main differences between his own critical common-sensism and Reid's philosophy of common sense (Subsec. 2.2.1). For the sake of completenss, take into account that Reid (1764, Ch. 5, Sec. 7) considered first principles as necessary and part of our own constitution so that "reason can neither make nor destroy them". In this way they possess a high degree of certitude like the existence of a material world does (1764, Ch. 5, Sec. 8).

[30] Suárez 1597, Disp. 31.

[31] Fabro 1941b, Chs. 2-3.

(Subsec. 2.1.2). This also implies that our knowledge is never adequate to the object but, in our effort of grasping antecedent cognitive ends addressed towards the object, knowledge determines subsequent cognitive ends that are more and more adequate to the object[32]. It is here that the mentioned theological understanding helps us to avoid a reification of the ideas we form as it was still the case for ancient Platonism[33] (Sec. 4.3), but allows us to attribute a kind of reality to them although not an ultimate one. Most of the misunderstandings among the philosophical schools arise from the difficulty in grasping the proper scope (and therefore the meaning) of the results that have been found, if any. Results always possess an objective value (and this is the objective side of formal constructs) but philosophy has the tendency to over-attribute significance to them and therefore also to lose the sense of their limitedness and partiality (which constitutes their provisionality).

Therefore, although limited our grasping of the formal constructs (logical axioms, mathematical theorems, principles and general laws of nature, and so on) can be seen as an intellectual achievement of creation, that is, the way in which matter (already informed from the beginning by intelligibility) elevates itself in a rational and reflected way to the understanding of creation and of its dependence on the ultimate Being. We may say that knowledge is a piece of ontology, especially when the world is understood not only from the perspective of the intelligible emanation but also as participatory (Subsec. 3.3.4). This is the philosophical, more specifically metaphysical, perspective that adds a value to the scientific and theological ones, an issue that has already been raised by Heidegger (Subsec. 2.3.7 and Sec. 3.4): Questioning and inquiring is being in its openness to being, being that is realizing itself through inquiry to knowing that – through knowing – it may come to loving[34].

Many overlook the human effort of self-elevation that enlightens the whole process of growing and dynamical complexification of the universe (through which new forms are continuously generated), underestimating that meaning is conferred to the latter precisely through our act of intellectual grasping[35]. In other words, by spiritually establishing a community with our Creator, we elevate ourselves to be conscious partakers of His providential plan for creation, thus con-

---

[32] Maréchal 1926, pp. 217-36.

[33] See the examination of this problem in Averroes, *AM*, pp. 73-77 (j77-j81).

[34] Lonergan 1963, p. 309.

[35] Following the Evangelical Anglican Thomas Scott, J. H. Newman said that growth is the only evidence of life (see Gilley 2009).

tributing to creation's manifesting its intelligibility in a reflected way. It is as if the stars would have a look at God's mind through us knowing them[36]. This intellectual adventure of the human self-elevation has brought some scholars to speak of an education of the human species[37], and this is something that we deeply share with the Enlightenment's idea of progress.

The previous considerations allow us to settle finally the issue of the activity of the mind (the subjective counterpart of Einstein's problem) in a proper context. The problem was not fully unknown to Middle-Age scholars. Although St Thomas clarified that humans cannot be directly imbued of God's ideas, he, following Avicenna[38], nonetheless stressed that perceptions are rather the trigger of a fundamental activity of the mind rather than a passive recipient of impressions[39]. However, if a general philosophical approach focussed on mind is to be recognized (although not yet translated in a true scientific strategy at that time), it is obviously the modern Idealistic one. Modern Idealism, at least in Fichte's original formulation, regarded mind essentially as an *act*, conceiving it, in the first place, as posing (*setzen*) its own contents[40]. We may suggest that idealistic philosophy in general[41] can help us very much to understand the mind as an *activity* that is able to bridge between physical items and purely formal constraints at a new level of complexity, i.e. in the framework of a reflexive knowledge (according to our examination so far). This is possible, however, only provided that we understand the specific character of the mind's activity without incurring in Idealists' extreme positions that imply the irrelevance of the domain of natural phenomena (Subsec. 3.1.1) or even make the rational conclusions of mind fully dependent on its own activity (Sec. 3.4).[42] Another important point to consider here is that any new research strategy cannot completely wash out previous methodologies and results (Sec. 1.2). The way in which philosophy deals with this problem is to integrate or try to integrate the relative philosophical approaches. We emphasize that Idealism, especially when taking into account the lessons of the other two schools, could became the philosophical framework for a research strategy centered on a new dynamic understanding of the whole of reality (Subsec. 3.3.3) and whose investingation core

---

[36] As lucidly forecast by Hamilton (1859, I, Lect. II, pp. 37-38) the only alternative is nihilism.

[37] Lessing 1777.

[38] Avicenna *de An.*, IV, Ch. 1; V, Chs. 5-6.

[39] Aquinas *S. Th.*, I, q. 78, a. 4 and q. 84, a. 6. See Fabro 1941b, Ch. 4.

[40] Fichte 1794.

[41] See Maréchal 1926.

[42] Hegel 1807; 1833. Gentile 1913.

will be humanity and especially the human mind. This also clarifies the circumstance that we are not denying the value of other philosophical schools, but are singling out those that have given rise not only to single approaches or methodologies but to long-run research strategies.

In conclusion, we consider it to be both possible and fruitful to see each of the three main philosophical approaches of our culture as basically dealing with one of the three "worlds" delineated in the general model proposed in recent studies[43], namely: (1) the physical world (Aristotelism); (2) the world of formal constructs (Platonism); (3) the mental world (Idealism).[44]

5.3 Some Current Issues

Having established some general aspects of the mental activity and its relations with the physical and the formal worlds, in this section we would like to deal with two particular topics that have arisen recently: the anthropic principle and the movement of transhumanism. The former concerns mainly the origin of humankind, the latter mainly its destination. These two topics will help us to deepen the issue of the mental activity that will be the object of the last section of this chapter.

5.3.1 *The Anthropic Principle*

A scientific point of view, raising several philosophical and theological issues, is the so-called "anthropic principle" which is concerned with the explanation of the existence of the human being (or in general of life as we know it) in the context of cosmic evolution: as a matter of fact, the basic parameters of the universe as they can be grasped by current physics and cosmology display those fine-tuned values that are required for life and intelligence to emerge[45]. It is really surprising to consider that these values must also be in harmony and none of them could be different[46]. We note that this perspective provides a scientific-based argument for the emergence of humanity, thus setting aside mythological narratives or erratic speculations claiming for its sudden appearance. Obviously, the anthropic principle

---

[43] Eccles 1970; Popper–Eccles 1977; Penrose 2004; Auletta 2011a, Ch. 25.

[44] We remark that to a certain extent Aristotle was aware of these three domains, that of physics, mathematics and of changeless substances (objects of metaphysics): see Sec. 3.4. In the latter case, he thought primarily about God.

[45] Barrow–Tipler 1986.

[46] Rees 1999.

has also been the object of many discussions, and different formulations and interpretations have been proposed. Without entering into the details of the different versions of the principle, we limit ourselves to the remark that certain weak formulations of the anthropic principle may be definitively too weak, up to the point of resulting in a truism: if one states that, because of the undeniable fact that we are here contemplating the universe, the cosmological and physical parameters have actually allowed us to contemplate it, then it is not easy to see what the consequences are. On the other hand, some strong formulations of the anthropic principle may imply an anthropocentric interpretation of the cosmos: one may assume that the currently known parameters of the universe are finely tuned not only as justifying the general order of the universe, but also as bringing about the eventual appearance of intelligent life as we know it (i.e. of the *Homo sapiens*), from the outset.

Thus, in order to evaluate the anthropic principle, we suggest the following statements to be taken into account[47]:

a) The universe is a unity and not a cluster of disconnected parts,
b) We need to consider the arising of life and humanity in the context of the whole history of the cosmos,
c) The conditions of possibility for this arising have been there from the start.

God, indeed, has created this universe as a whole. This vision grasps, from a theological point of view, what we have philosophically said about convergences (Sec. 3.3). Indeed, precisely those solutions that have been spontaneously selected from the outset at a pure physical level are the same that allow for life and intelligence according to the thesis of emergent monism. In Subsec. 3.3.6, we have stressed that our universe goes through a growing process of complexification. When a more complex reality emerges from a less complex one, the immediate result is that we have a larger number of components or factors that are integrated in the new system (Subsec. 3.2.2). Clear evidence for that is the higher number of elements and interrelations in a bio-molecule like a protein relative to any a-biotic molecule such as water. So, obtaining the more complex from the less complex implies a growing number of constraints and therefore also a lower probability of finding the ordered or stable configuration. In other words, the number of possible disordered (and unstable) configurations grows exponentially

---

[47] Tanzella-Nitti 2009, pp. 74-75. See also Edwards 2010, pp. 39-43.

with complexification. Such a consequence is evident by taking into account the entropic considerations we have developed in Subsec. 3.2.4 and also implies, for instance, that all the different bio-molecules share much more constraints among them than all a-biotic ones do. This explains that along the process of growing complexification there is also a sort of growing *canalization*. This statement, however, should not be taken in the sense that we have less and less variety (this would be in contradiction with the expansion of the space of all possible configurations), rather in the sense that we have not only more and more shared constraints but also new properties and capabilities[48]. Therefore, at any level of complexity in our world, we have an emergent variety of configurations that also represents the conditions of possibility of further and higher levels of organization. For instance, life would not arise without the variety of bio-molecules. As already mentioned, this is what Aristotle called the material cause (Subsec. 3.3.7). However, since high levels of complexity also imply a high number of shared constraints among the systems at that level of complexity, this also implies that the general characters determining the high intelligence of humans (like making use of symbolic systems, being organized in political societies, providing theories about our world and therefore interpreting it, being endowed with intentionality, and so on) are likely to be *universal*, so that we should find them in every being of the universe to which we reasonably attribute rationality. However, the specific characters of humanity that depend on the *contingent* history of evolution on Earth cannot be considered as universal as such, and mixing these two very different issues would be a major theological and philosophical mistake. Such an explanation integrates what we have said in Sec. 5.2 about the intellectual self-elevation of humanity, since it considers the latter as a part of a universal dynamical process of complexification in which the continuous emergence of new forms and constraints represents a necessary departure point for this intellectual journey. St. Thomas has stressed several times that there is a certain continuity between humans and other animals. This is particularly evident when he speaks of animals understanding ends[49]. This is also in accordance with St Thomas' statement that the human intelligence reaches intelligible notions by abstracting species from material conditions[50].

---

[48] Auletta 2011a, Ch. 6.

[49] Aquinas, in *S. Th.*, II.I, q. 6, a. 2, says: "Imperfecta cognitio finis invenitur in brutis animalibus", and concludes: "Soli rationali naturae competit voluntarium secundum rationem perfectam; sed secundum rationem imperfectam, competit etiam brutis animalibus". And in *S. Th.*, II.I, q. 12, a. 5: "Etiam bruta animalia intendunt finem".

[50] Aquinas, *S. Th.*, I, q. 79, a. 3. See also *S. Th.*, I, q. 85, a. 1.

The question is therefore: Does the anthropic principle deal with humanity in particular or with intelligent beings in general? If it deals with humanity as a particular biological species, we would fall again into strong formulation. Indeed, the idea that all the parameters of the cosmos have been orchestrated to bring *Homo sapiens* to existence (that is, the intelligent being living in our planet and descending from an ape-like ancestor together with all its physical characters) seems to be a curious mix of the absurd and arrogant. We believe, on the contrary, that the anthropic principle should deal with the rational being in general, that is, with those features that make humans really able to give rise to a rational relation with the physical and formal worlds (knowledge), as well as an ethical and emotional relation with other human beings; moreover, all of this also encompasses a spiritual relation and religious covenant with the Creator. It is here that a correct understanding of the principle can become very helpful. We are indeed able to establish a relation with God precisely because we are able to transcend our biological dimension, and this constitutes the *universal* spiritual value of humanity. Therefore, we stress that the rational being (*capax Dei*) does not define a biological species (it does not coincide with *Homo sapiens*) but is a *genus*: any being capable of thinking symbolically, of having a culture, interpreting information, searching for a meaning, should be credited as a rational being and therefore as a *Homo* in this enlarged sense, without obviously denying a sort of "general" animal dimension of humanity (Sec. 5.1). Some of our own ancestors or parallel branches of human evolution also pertained to this genus, even when they do not pertain to our biological species (i.e. when we could not interbreed with them). Humanity should be certainly attributed to the *Homo Neanderthalensis* who buried his dead and probably had a form of religious cult, but also to *Homo erectus* or *Homo ergaster* as well, although it is quite difficult to judge about their intellectual level due to the huge time span dividing our species. Similar considerations could also apply to our own descendants as well as to other intelligent beings who could live somewhere else in the universe[51].

Theology should take into account such considerations. Let us recall that Christ's Incarnation unifies an eternal aspect and a temporal happening. As far as L. Ladaria reminds us, according to Duns Scot[52] Christ has been eternally predes-

---

[51] Tanzella-Nitti 2010.

[52] Duns Scot *Rep. Par.*, b. 3, d. 7, q. 4. See also Ladaria 1995, pp. 45-46 and Tanzella-Nitti 2009, pp. 83-86 and 108-112. Such an idea was already supported in the 12th century by Guillelmus a Sancto Theodorico, *Exp. Ro.*, lib. 1, 61-79.

tined to Incarnation and this has a universal cosmological value that goes far beyond the historical advent of Jesus (without diminishing the redemptive significance of this event for humanity, Sec. 4.5), therefore preceding Creation and even constituting the reason and ultimate goal of it. This aspect has also been recently recalled by Pope Benedict XVI.[53] It is, however, an issue that goes well beyond our competences and the scope of this book.

5.3.2 *Transhumanism*

An integrating cognitive enterprise is also called to elaborate on future technological developments. Nowadays, the implementation of technological devices and procedures in order to artificially enhance several aspects of our physical, biological and psychological components, is advocated by the proponents of a quite diffused movement known by the name of *transhumanism*. Not all the proponents of transhumanism maintain the same worldview; the solutions too, which have been envisaged or emphasized with regards to a number of concrete problems, have many variations. A shared programmatic tenet, however, stems from the assumption that human nature is still a "work in progress" so that a pervasive employment of technological means is sought in order to overcome current human limitations[54]. Does this option represent a real chance to better develop our faculties, overall lifestyle, spirituality, and with that the peaceful living among populations as well? Or, is it merely a presumptuous attempt at altering the course of natural evolution and interfering in an uncontrolled way with complex biological and mental processes by technological means? Might it help the flourishing of a new mature humanism or will it lead us to a mortification of the human condition enslaving people with devices and prostheses? Will it promote social welfare or produce new social disparities and tensions?

We think that these issues deserve a careful reflection and that at the present stage we should try to assess the general consequences of the employment of new and still unsatisfactorily tested technologies, as well as to evaluate possible enhancing interventions case by case[55]. Whenever technology makes it possible for new radical

---

[53] In his general audicence of July the 7th, 2010, he said: Duns Scotus makes it clear that the Incarnation is the greatest and most sublime work of the history of salvation, and that it is not conditioned by any contingent circumstance.

[54] See www.humanityplus.org.

[55] Bostrom–Sandberg 2008.

changes, new intellectual problems suddenly arise that cannot be easily solved at the beginning of such a transformation. Consequently, transitory situations are induced in which new ethical and juridical standards are still to be introduced, so that a certain caution is demanded. Indeed, in the near future, direct applications to the human subject of artificial devices might modify our own condition somehow, for instance by extending the range of our perceptions, by increasing our learning capability, by delaying our aging, or – more imaginatively – by connecting our brain or our conscious state to a computer. It is difficult, however, to imagine that these interventions will significantly enhance typical human capabilities like having insights or developing an aesthetic taste that are even crucial for science, since these higher human capabilities are the result of very complex and interactive processes[56]. Indeed, only basic intellectual capabilities are strongly modularized and could eventually be enhanced by technology. On the other hand, one should also consider, among many other factors, the attainable benefits in the treatment of physical malformations, dysfunctions as well as in healing old and new diseases.

A relevant aspect is the following: a human society cannot exist without a culture, and we have stressed how specific cultural traditions are – of which religious beliefs constitute a very important part world-wide (Subsec. 2.3.1). Now, the recent progresses in the exchanges among human populations may lead to the danger of substituting a globalization process to the diversification of cultures. In other words, science and especially technology (with a "religion" of consumption) claim to substitute such a diversification. However, technology (and even science) cannot substitute culture (Sec. 1.1), the risk being an immense spiritual (and cultural) impoverishment of humanity[57].

One of the main problems raised by transhumanism is the issue of the life span of humans. Future technologies could allow for a considerable prolongation of human life. Some of the most radical representatives of this movement even believe that humans could become immortal. This is obviously very naïve but also raises the huge problem of the essential desire of humans to endure forever, which is also basic to the religious feeling (Subsec. 2.3.1). The most dramatic condition of humanity is the felt contradiction between a mind that longs for universality and a body that is necessarily limited. Perhaps not by chance, Schrödinger considered dealing with time, and in particular the effort to overcome temporal lim-

---

[56] Auletta *et al.* 2011b.
[57] As recalled by Benedict XVI (see Sorge 2006, pp. 96-97).

itations, as *the* theological question as such[58]. Since the Christian message deals with the resurrection of the whole person (that is, of the bodily persons, raprespenting a transfiguration of both the person and the whole Creation[59]), we expect that God's Kingdom will display an enhancement and ennoblement of the dimension of matter and not its negation or annihilation.

For this reason, future technologies should be also devoted to solve our ecological problems. We remark that the only way to avoid the risk of the overexploiting and consumption of the Earth's resources is to search for responsible development and use of new-generation technologies. Moreover, in a theological framework, we should take into consideration that these developments may eventually allow for the performance of our duty of stewardship towards the created nature in a beneficial way that, until yesterday, was still unconceivable[60]. This stewardship can only be realized considering the perspective of the new Creation (the second coming of Jesus and the resurrection of the dead: Subsec. 2.3.6), a perspective that transcends but includes what we call history, nature or science[61]. The supernatural mission of the Church does not separate it from history and temporal realities in which this is rather incarnated, which means that the Redemption through Christ also comprehends the temporal order which it is called to perfect with evangelic spirit[62]. This also makes the Gospel message really universal since addressed to everybody indipendently of his/her faith. This context is the framework of the preferential option of the Catholic Church for the poors[63]. In other words, God's Kingdom should not be understood as a denial of the first (current) Creation in a kind of ontological dualism (earth and matter are bad and heaven is good) but in terms of an eschatological duality between the present age (in which Christ has incarnated) and the age to come (in which Christ will come back).[64] Indeed, the resurrection will be of our bodies. For this reason, St Thomas says that the intellect is the form of the human body[65] to which it is united not acciden-

---

[58] Schrödinger 1958, p. 149.

[59] Wright 2007, pp. 52-63, 104-119, and 159-176.

[60] Message of Benedict XVI for the celebration of the World Day of Peace, January the 1st, 2010 (http://www.vatican.va/holy_father/benedict_xvi/messages/peace/documents/hf_ben-xvi_mes_20091208_xliii-world-day-peace_en.html). See also Heap-Comim 2008.

[61] Wright 2007, pp. 83 and 214-217.

[62] Sorge 2006, pp. 18-19 and 145-47.

[63] Sorge 2006, p. 88.

[64] Wright 2007, pp. 105-106. See also Edwards 2010, pp. 24-33 and 91-166.

[65] Aquinas, *S. Th.*, I, q. 76, a. 1.

tally[66], and that the perfection of the body is required for beatitude[67]. This also implies that we cannot work in this perspective presuming to be the masters of creation. We only work for the Kingdom of God to come but we cannot determine or predetermine in any way this process and not even know whether or not we are the only intelligent beings involved, whether we shall succeed or not[68]. This is ultimately the business of God to which we are called to participate (Sec. 4.5), but without the presumption and arrogance to elevate any of our ideas, ideologies, parties, society, group or whatever form an Utopia may take, to be the door or the leverage for the Kingdom of God (Subsec. 2.3.4). This demands a great modesty but also a great sense of responsibility, implying a deep engagement from both a moral and a religious point of view[69], which however should not be separated from our search for Truth[70]. This also implies that there is not continuity or a progress starting from any of the aspects of this world and guiding us to the new Creation. This explains why the conflation of God's Providence with an idea of finality ruling the first creation (and therefore giving a certain direction of march) clashes with the core of our religion (Subsec. 4.1.3 and Sec. 4.3). These semplifications do not take seriously into account the dimension of evil in all its forms (moral, physical and metaphysical) resulting in a fundamental inability to cope with this problem in a way that is simultaneously propositive and realist.

Therefore, the movement of transhumanism in its general conception allows us to think in modern terms about a crucial philosophical and theological issue concerning the human being, namely, the capability of *self-transcendence*. Does self-transcendence mean going beyond natural limits as an instance of man's arbitrary will and autarchic self-determination? Or is it possible to understand it in a more profound way, consistently with the limits and faculties characterizing humanity but trying to overcome the actual human condition? Shall we human be able to transform what is disorder and garbage in new source of order and creativity? What is hate and conflict in establishing new, higher forms of community? As the Psalm (117 (118), 22) says and Jesus recalls (Matthew 21, 42): "The stone which the builders have rejected, this has become the head of the corner". In the previous subsection, we suggested that our possible establishing of a relationship

---

[66] Aquinas, *S. Th.*, I, q. 76, a. 6.

[67] Aquinas, *S. Th.*, II.I, q. 4, a. 6.

[68] Wright 2007, pp. 156-158.

[69] Wright 2007, pp. 218-224 and 228-229.

[70] Benedict XVI, *Caritas in Veritate*. See also Sorge 2006, p. 98.

with God stems from the capability that we have to transcend the biological dimension. Self-transcendence involves the constitutive openness of the finite and in particular of the human being towards the divine transcendent reality, to the possibility of recognizing it as independent from anyone's will or intellectual constructs, and to the faculty of freely acting in adhesion to it[71]. We have already mentioned (Sec. 4.5) that Einstein, although denying a personal dimension of the divine, considered God as super-personal. It is worth noting that Einstein regarded religion as that dimension that liberates humans from selfish desires and thoughts, and opens them to aspirations that have a super-personal value. A religious person is devout in the sense that he/she has no doubt about the significance and loftiness of those supernatural objects and goals like absolute truths and values which neither require, nor are capable, of rational foundations since they exist with the same necessity and matter-of-factness as he to himself[72].

5.4 Self Canalization

Within the general cognitive interest of the present work, we may say that whenever one has to admit that the results of an intellectual research are not depending solely on human efforts, that the natural world's features encountered or discovered show universal validity, one also acknowledges (at least implicitly) the limit of our knowledge. Therefore, it is possible to experience an access to a dimension of truth that is not fully grasped or merely achieved, rather basically *received* though never in an exhaustive way (Sec. 5.2). In other words, the results of an investigation that is freely undertaken show their autonomous validity with respect to the investigators, who therefore may recognize knowledge (and themselves) as being ultimately founded on something else, which is in a strong sense ulterior and not totally knowable, but only partially perceived during one's cognitive journey. Such an apparently paradoxical state of affairs may be seen as the virtuous circle through which human beings progressively realize their very own nature: in longing for universality, we transcend our biological constitution and individual perspective, yet we discover the Truth that we were looking for as something that we cannot own but that we have to accept and to which we can participate in respecting our limits.

In the light of the previous considerations, it is useful to come back to the philosophical issue of mind. Some Idealists ended up in conceiving of an absolute self-

---

[71] Rahner 1976. See also Edwards 2010, pp. 43-45.
[72] Eisntsein 1939, pp. 22-23.

determination as the proper act of the human mind in an all-encompassing subjectivity that does not acknowledge any otherness (Secs. 3.4 and 5.2). We may say that precisely the denying of any sort of irreducible otherness makes such a conception incapable of realizing the relevance of actual relationships of the mental world both with the formal and the physical ones, as well as with the intersubjective dimension. Let us consider these three aspects separtaely. We are aware that this mental (and symbolic) activity is triadic by definition (always involving the three worlds), but it can be very helpful to put the accent on dyadic relations mind-world, mind-forms, mind-minds.

5.4.1 *The Mental and the Physical*

In the present chapter, we argued that one of the major problems that a new research strategy should face is the connection between anthropology and naturalism (Sec. 5.1). On the one hand, we can envisage the human mind as an emergent process from biological evolution according to emergent monism (Subsec. 3.3.5). This means that the mind cannot be a sort of detached substance, or a "super-instance" relative to the brain. Indeed, brain processes are *not* generated by the mind: We can even say that most of our operations are performed in a spontaneous way by our brain without any rational planning or conscious intervention[73]. What the mind can do, instead, is to establish rational plans, concepts, values, and so on, as *general frameworks* in which those spontaneous operations take their course[74]. In other words, the mind rather than generating such processes can impose certain limitations and even block them thus redirecting their spontaneous course[75]. The way in which the mind is able to canalize the brain operations is by resorting to physical events occurring outside the brain and by treating them in informational terms: vocal emissions, gestures, physical objects, or any other kind of structure or item can be combined by the mind in pure physical terms but according to syntactic rules that are not physical themselves. This means that the mind throughout the whole of the person's brain and body takes an active part to the dynamical interactions occurring in the physical world to which the person pertains as a material entity[76] and uses this as a way of conditioning its own neural processes.

---

[73] Jeannerod 2006; 2009.

[74] Auletta 2011a, Sec. 18.5 and Chs. 20 and 24.

[75] Libet 2004.

[76] Von Helmholtz (1867, pp. 586-611) had great historical merit when he stressed the active role of the mind at the level of perception.

However, this is achieved without violating the closure of physical laws (Subsec. 2.1.6) and by having rather a constraining influence on the physical dimension. As understood by M. Merleau-Ponty, humans, as all organisms, are not like instruments that passively respond to external stimuli but rather they actively participate in the constitution of their experience: this experience could not exist at all without the organism's motion and activity[77]. Ordinarily, one goes from a nebula of impressions and sparse sensations to a more determined perception, since perception is a solution to a previous problem. Therefore, association is never an efficient cause that provokes the response but it only makes an intention probable or attractive. Also the similitude with past perceptions is not active without a certain form given by the present perception. The past does not act on the present but is rather evoked by the present mental activity (Subsec. 3.2.5). This means that all things are always perceived on a background and any percept is constituted in a context. In particular, at the human level, the text of the world is not copied by our senses but is rather dynamically-symbolically integrated and constituted. Consequently, perceptions are always given in a horizon of meaning whose ultimate root is not only in the symbolic activity of the single mind but rather in the society of minds and their intentionally shared meanings, as we shall say below[78]. The first operation of attention is to build a perceptive field that one can dominate. These fields are new regions in the total symbolic world built through categorical and conceptual patches. Any new acquisition allows for a new integration of the previous facts. While a physical system only adapts to external conditions, the organism poses by *itself* the terms of its vital problem[79]. In other words, whilst experience is essentially equivocal, human symbolic activity is integrative. All the parts of the body are actively integrated and evaluated in programming and planning, in establishing values and rules, and in developing the symbolic activity in general.

In this way, the mind, having a constraining influence on the brain and the body, through its participation in the dynamism of the physical world is thus able to operate a process of *self-canalization*: the human mind is both the highest result so far of a process of growing complexification and canalization of nature, and the way for exerting a new, original, *active* canalization of the physical world. This is also the manner in which a new connection between the physical and the mental

---

[77] Merleau-Ponty 1942.
[78] Merleau-Ponty 1945, pp. 9-77.
[79] Merleau-Ponty 1945, pp. 81-232.

worlds is established as well as a connection between the formal and the mental dimensions is realized. Let us consider the latter point.

5.4.2 *The Mental and the Formal*

The mind is also able to establish a unique connection with the formal world. We have already seen how many empirical and formal constraints there are on our knowledge (Sec. 5.2), but we have not so much emphasized the process that brings us to such results. We recall that the way in which we humans arrive at certain conclusions is contingent, but when particular results of general validity have been found, we are no longer in the position to change something in these formal constructs according to our arbitrary will. It is true that we can excogitate new classes of objects that do not obey acknowledged laws or possess unexpected characters (Subsecs. 2.1.3 and 3.1.2). However, as mentioned, this or that new class of objects also exerts certain constraints (simply due to the fact that these objects share certain properties), so that the game goes on. Yet, it may still be legitimate to ask: what forbids us to think that it is precisely this never-ending process that shows that contingency is the last word, and that the constraints we always find are themselves contingent and provisional? In a word, the fact that, in this game, our understanding grows, science continues, new formal aspects and new elements of reality are revealed. And this knowledge grows because the results found on the path have an objective value and can therefore be integrated into further knowledge not only by ourselves but also by others: only in this sense are they true acquisitions, although this integration cannot be understood in terms of positive accumulation only but as a continuous process of self-correction and updating (Subsecs. 2.1.2 and 2.1.4). Moreover, there is no necessity to reach those particular results and we shall certainly never reach many others, at least in any finite time span. We can also assume that if other forms of intelligence existed, they could reach some of those other results and even follow alternative paths of investigation. However, it is likely that we would cross their research journey at some point and we would certainly cross it in the long run; this is a form of convergence. Moreover, the paths may be different but they represent the form of inferential-hypothetical reasoning that is the way in which the mind accommodates to the relations that exist between cognitive principles and results, between universals and particulars (experiences). Therefore, our results are both posed (*setzen*) by ourselves, as Fichte said (Sec. 5.2), and objective at the same time. They are posed by ourselves because they are sort of milestones on our never-ending research path but are objective as well, since they are sort of anchoring points for our further investigation, or ladders – if you will – for climbing further and further; ladders, moreover, that could be potentially used by others.

It is important to consider carefully the issue of conceptual posing. When Fichte introduced this terminology[80], he was fully aware that any tautology of the kind "A is A" has a conditional form (When A, then A) and that only such a hypothetical reasoning is posed as such. The term A (which represents an arbitrary content) is not posed as such, or at least we do not know how and when it was posed since it may be itself the consequence of a previous inferential process[81]. Therefore, this conditional form represents a nexus between an unknown position of A (expressed by the antecedent "When A") and an absolute position of A through this conditional intellectual act (expressed by the consequent "then A"). In this way, A becomes posed in the "I" precisely in the same way as the conditional form already was, and the "I" becomes conscious of its activity as such[82]. Thus, it becomes clear that Fichte's act of position is not reducible to a subjective introspection or apperception (Sec. 3.4) because it has a universal validity for any finite mind (which justifies the term "I" in all its generality).[83] There are also many different states or processes empirically occurring in the mind that do not have the character pointed out by Fichte. However, this act of posing cannot be ascribed to an absolute and impersonal Being either, because it has the fundamental marks of the mental activity. In other words, although the hypothetical act of reasoning and therefore the posing of the conceptual contents is an expression of human freedom and creativity, the starting inputs as well as the possible conclusions to which we may arrive at are not dependent on this act itself: the first ones are in general given through our experience, the latter ones are acknowledged to be universal, certainly

---

[80] Fichte 1794, § 1, 1-5. In the following we resume his original derivation. See also Lonergan 1957, pp. 344-346.

[81] See also Fabro 1941a, Introduction, and 1941b, Ch. 1.

[82] See also Hamilton 1859, I, Lect. XVI, pp. 291-292. In the words of Lonergan (1957, p. 353) a contingency becomes conditionally necessary.

[83] Peirce (1905, p. 348) says that when we are concious of a belief that has been determined by another given belief, but we are not conscious that it proceeds on any general principle, as it is the case for *Cogito ergo sum*, then we do not have a true reasoning but an acritical inference (which is however a widely used and even necessary form of thinking). The attempt of Fichte to subtract the analysis of the mental act to a personal apperception that still affected Descartes' (or St Augustine's) analysis is precisely a way to restate the mental act in terms of general principles.

as a result of an investigation, but precisely as a result[84]. In the subsequent developments of Idealism and also of Fichte's own philosophy this point of view got lost. Indeed, when we consider Fichte's later work[85] it is affirmed that the absolute Being (which here also takes the impersonal character of an absolute reason: see Sec. 4.5) cannot be posed by something else but must be posed by itself, and in this way it becomes an absolute self-construction that recalls the Spinozian concept of *causa sui*[86]. It is also possible that there is a confusion here between the mind's *activity* and God as pure act, self-thinking thought, as understood in the Aristotelian tradition (we recall that God's mind cannot be discursive, Subsec. 4.1.1). The consequence is that the delicate balance between objectivity and provisionality of the *act of posing* is destroyed, since the peculiarity of the finite mind's activity consists in rationally founding a content that is not self-given. This turn in Fichte's positions did not prevent Hegel from relegating the act of *Setzen* to a mere moment of its *Logic*, since it still had the flavour of a subjectivity: it indeed only characterizes the logic of *Wesen*, that is, of reflection[87]. Commenting on a crucial passage, an authoritative interpreter of Hegel explains that here *Setzen* is confined to a form of pretence (*Schein*).[88]

The view proposed here fits perfectly our critical realism and fallibilist epistemology that tries to avoid the dead ends of both conventionalism and foundationalism (Sec. 2.1), i.e. that knowledge is a projection that searches for, and grounds, new and more universal foundations. Summarizing: although the inferential path leading to certain truths is contingent and is what constitutes our finite dimension as creatures[89], the results or the conclusions of our investigation may not be. In this case we *assent* to them with the same certitude with which we assent to the empirical facts or personal experiences that have started our inquiry (although it

---

[84] For this reason, Leibniz (1765, pp. 347-48) says that there is an exaggeration in the Cartesian statement *Cogito ergo sum* since it is not only clear that I think but also that I have different thoughts: sometimes I think of A, sometimes of B. This is still more interesting considering that Leibniz is dealing here with primitive truths both of experience and reason (he acknowledges that the latter are essentially hypothetical).

[85] Fichte 1804, pp. 185-186 (SW 233-235).

[86] Spinoza 1677, I, prop. XI.

[87] Hegel 1833, I, p. 109 and II, pp. 16-20.

[88] Hegel 1833, II, p. 269; Theunissen 1980, pp. 424-425.

[89] But it also makes humans to be constitutionally involved in the search for truth: see Hamilton 1859, I, Lect. I, pp. 12-13.

is a case of reflex assent and not an immediate or simple one).[90] If assent is the acceptance of truth, and truth is the proper but final object of the intellect, and as no one can hold conditionally what by the same act he holds to be true, then assent is an adhesion without reserve or doubt to the proposition to which it is given[91]. *Inference* is always inference; even if demonstrative, it is still conditional; it establishes an incontrovertible conclusion on the condition of incontrovertible premises or it exhibits a sufficient reason for the conclusion. To the *conclusion* thus drawn, assent gives its absolute acknowledgement. In the words of Sir W. Hamilton: reason, concatenating thoughts and objects into a system, and tending always upwards from particular facts to general laws, and from general laws to universal principles, is never satisfied in its ascent till it comprehends (which in fact, it can never do) all laws into a single formula, and consummate all conditional knowledge in the unity of unconditional existence[92]. The highest level of this acknowledgement, following Hamilton's reception of scholasticism, can be said to be intuitive or immediate knowledge in opposition to the abstractive, representational or mediated knowledge[93]. This also establishes a certain continuity between what we can call assent to empirical facts, objects or belief in their existence and what we can call true faith, that is, in God. Indeed, both have a common root, although in the former case the assent is a consequence of knowledge, whilst in the latter case assent and knowledge are *ex aequo,* because here the voluntary component is more important[94].

---

[90] This is why Reid stressed that we have an immediate certitude of the existence of material objects (see the insightful commentary in Hamilton 1859, II, Lect. XXIV, pp. 86-115). Peirce (1905, p. 348) says that our enquiry cannot be beginningless or endless but starts with perceptual judgements and ends with beliefs of a general and recurrent kind. Compare this with what was said in Subsec. 2.1.4.

[91] Newman 1870, Ch. 6. See also De Morgan 1847, pp. 191-195.

[92] Hamilton 1859, I, Lect. IV, p. 68. If we could actually reach this final end of knowledge, the latter would be no longer discursive.

[93] Hamilton 1859, II, Lect. XXIII, pp. 68-72. This seems to be also Aristotle's own point of view (*An. Post.,* 72b18-25), only that this applies to any principle that we may assume, even provisionally. So, we do not need to take the term "first principles" in any absolute sense. It is possible that Aristotle would agree also on the latter point (see for instance 76a16-25). On the other hand, it is important to stress that this intellectual intuition (called *nous* in 100b12) is not a methodology for reaching the truth (which on the contrary is an inferential process that combines natural deduction, abduction and induction) but a state of the mind (that we have called insight) that we reach once we have found the searched solution to the intellectual problem from which we started (on all this matters see Anagnostopoulos 2009b).

[94] Aquinas, *De ver.,* q. 14, a. 1.

We have said that the mind tries to rationally found its contents. Leibniz[95] distinguished between the principle of sufficient reason, dealing with the conditional attribution of something to something else and covering any possible *content* of our thoughts, and the principle of identity covering tautologies only and therefore expressing the logical *form* of thinking. Lonergan[96] identifies the sufficient reason with the highest level of consciousness, namely the rational one: the law of sufficient reason is a single law of outmost generality and expresses the refusal to assent unreservedly on any lesser ground. Since intelligence looks for intelligibility[97], we may perhaps consider such a principle as the most general methodological principle (cutting across any of the cognitive fields involved here) guiding our search for intelligibility. It is important, however, to understand the principle of sufficient reason not in the way in which Leibniz understood it (Everything has a reason to be as it is[98]), which still presupposes a necessity of the antecedence, but in this sense (Subsecs. 3.3.1 and 3.3.4):

Of everything a ground may be found that enables a result that we hold to be true.

In this way the principle of sufficient reason becomes a reformulation of Fichte's act of posing. Moreover, this formulation of the principle covers both anterograde (deductions) and retrograde (abductions and inductions) inferences (Subsec. 3.1.2).[99]

In other words, the principle of sufficient reason means that the essence of knowledge is to rationally ground some initial facts of the experience or immature assumptions[100].

---

[95] Leibniz, 1712-1714, §§ 33-36. On this issue, see also Hamilton 1859, III, Lect. XIII, pp. 237-239; consider, however, that we use here the terms hypothetical and conditional as synonyms.

[96] Lonergan 1957, p. 346.

[97] Lonergan 1957, p. 347.

[98] Leibniz 1710, I, § 44; 1712-1714, § 32. He often calls this the principle of the determining reason. In the words of C. Wolff, 1730, § 70: Nihil est sine ratione sufficiente, cur potius sit, quam non sit, hoc est, si aliquid esse ponitur, ponendum etiam est aliquid, unde intelligitur, cur idem potius sit, quam non sit.

[99] In Auletta 2006, the principle of sufficient reason was criticized since it was interpreted in strict Leibnizian terms. In the present, generalized formulation is in fact to be accepted as it is also able to account for scientific findings like those related to quantum physics. See also Auletta 2011d.

[100] In this context, we remark that the German word *Grund* (often used by Leibniz and Wolff but also by German Idealists), which is currently translated as *reason*, also means *ground, foundation*.

Summing up, self-canalization is the process through which the mind *freely* acknowledges and accepts the *necessity* of concepts, principles and values, that are *not* the result of human creation, even if the path (the hypothetical reasoning stimulated by our own experience) leading to them is the most incredible witness of human creativity[101]. We hope that what has been said so far may represent a fruitful answer to Einstein's worry recalled in Sec. 5.2.

5.4.3 *Intersubjectivity*

Properly speaking, self-canalization is the activity through which both the connections between the mental and the physical world, as well as between the mental and the formal ones are strictly entrenched in the *intersubjective* dimension that constitutes the whole of human culture. Human culture essentially consists of a process of symbolic sharing and information interpretation through which we are able to canalize our own development (thanks to education and transmission of culture to the next generations) and, to a certain extent, even our own evolution (through our ability to reshape the environment and to give rise to new forms of artificial contexts playing the role of frameworks for our species, Subsec. 5.3.2). Then, culture is the ecological niche of our species or of our genus (Subsec. 3.3.1). Human beings are distinctively characterized by the fact that they try to establish reasonable (and often causal) connections between events that are distant in space and time[102]. Other primates, although able to understand quite well "objective" causal connections (that is, connections that do not depend on our action on other organisms, as it is the case for lower mammals), are limited to objects that are relative contiguous in space and time and therefore perceptible. Now, the crucial point is that such an effort to extract information that is hidden to our immediate experience cannot be successful if not by establishing socially shared interpretations, going from the academic sharing of scientific theories up to religious traditions contributing to give sense to a common cultural humus.

This clearly shows that, although conventional, culture is never arbitrary, being another manifestation of the fundamental connection between contingent paths of exploration and results endowed with universal validity that is the quintessence of humanity (Subsec. 2.3.1). This process of transmission is only possible when the new generations to be educated believe in the previous generations and their traditions in order to reach further, more pondered and grounded beliefs, so that

---

[101] Auletta 2011a, Ch 25.
[102] Auletta 2011a, Secs. 18.1 and 23.3.

the process of self-canalization can be seen as a dynamical integration of tradition and novelty starting from a lower belief and ending in a higher belief or also going from empirical, less rational assent to higher and socially shared rational assent.

The community of minds is established through relations that are ultimately of a social kind and is based on empathy or on the affective dimension in which both the individual and the community live in a process of universal spiritual communication. Empathy is the basis of human intentionality. Human beings are intrinsically characterized by intentionality, that is, the ability to share symbols (included symbolic intentions) towards a third party. When children learn to speak, they try to establish a visual contact with the mother and follow her gaze trying to share reference to the object of the mother's speech[103]. The "community" of the physical world is based instead on the universal dynamism in which the global and local are entrenched (Sec. 3.3). The "community" of the formal world is expressed by hypothetical-inferential relations between the universal and the particular.

Therefore, the mind allows, so to speak, for a flux between the other two separated worlds. In this sense, the arising of humanity as *capax Dei* represents an accomplishment of creation (Sec. 5.2), and our cognitive journey is a path leading to Truth, that is, a form of revelation[104], which can truly be called certitude[105], it is the ultimate convergence point of our intellectual and religious journeys[106]. We can never reach this point[107]. This is why it is here that knowledge becomes a sort of rational prayer ending finally in a true prayer, the way in which we are able to have a spiritual community with God, which in the words of Rosenzweig cannot be known (*Erkennen*) in the proper sense of the word, nor experienced (*Erleben*), but grasped (*Erfassen*) as all and one in such a communion[108].

---

[102] Auletta 2011a, Ch. 20.

[104] Hamilton 1859, I, Lect. V, p. 83.

[105] Newman 1870, Ch. 7.

[106] Mewlana (*SW*, 4, p. 23) says: "The many things of this world are a trial appointed by God, for they hide the single reality. There is a saying that the saint is one, humankind is hundred". And: "For the goal is neither going up or down. When the goal shows itself, it will be neither above nor below" (*SW*, 4, pp. 48-49). However, the veils represented by finite things exist for a purpose: we do not have the power to endure God's beauty directly and immediately (*SW*, 4, p. 59).

[107] Hamilton 1859, I, Lect. III, p. 60.

[108] Rosenzweig 1921, §§ 230 and 416.

6

CONCLUSIONS

Let us summarize in a few words the main results that we have found so far. We have pointed out that every cognitive enterprise aims ultimately at truth. However, we have also stressed that no discipline (not even theology) can fully possess such a truth. Nevertheless, there are relevant differences between science, philosophy and theology concerning the way in which they approach this problem. *Science* experiences a potential contradiction between its aspiration to universality (to a final truth about the universe) and its current practice that is specialized and focussed on particular results. Such a potential contradiction cannot be completely solved in the framework of science itself but needs further horizons of understanding. *Philosophy* indeed represents a further dimension of knowledge aiming at freeing science from the crystallization in specific results, methodologies, assumptions, paradigms and strategies. Since science is the field of investigation that is more connected with facts, there is always the temptation to consider the facts that we experience (which are always partial) and their received interpretation as ultimate. The job of philosophy is to show that the whole of science is finally hypothetical and that it can find its own foundations ahead, in a dynamical effort at further and further grasping reality. This means that philosophy is essentially a critical enterprise devoted to stimulating research towards the understanding of the general characters of both our world and our knowledge. Philosophy also experiences a certain dichotomy: one the one hand, it strives for universality and specifically for a universal dimension of rationality, which here we have called intelligibility; on the other, its critical method and especially the fact that it is far less connected with empirical facts than science is, makes the risk of falling into scepticism and relativism a very concrete one, where any universality and even rationality is ultimately lost. The problem is that philosophy cannot justify by itself the universal dimension of rationality that it is addressed to. This demands indeed a *rational faith* in the intelligibility of our world and therefore also in the possibility that our theories will ultimately agree with reality. The foundations of this rational faith can only be provided by *theology* and its ability to rationally argue that God's creation and government of the world is the ground of that intelligibility. The crucial point here is that rational theology cannot do this job properly without a deep connection with scientific results and empirical knowledge. A theology disconnected from the latter will be growingly acquainted with easy and wishful solutions, and will therefore fall back into superstition and mythology. It is indeed very appealing to find a meaning of the universe that is immediately connected with the providential plan of the Creator, but it is an immature and irrational way to express our need of protection. This would bring us to claim both to understand the details of that plan theologically that in fact remains a true mystery for us, and to try to enslave empirical research. On the contrary, we think that theology cannot determine scientific research strategy and programs even if it can and should constitute an important background.

Therefore, philosophy should be understood as the mediation instance between theology and natural sciences. The way in which philosophy can be helpful to the other two fields is to provide some general and abstract principles that can both find their root in crucial scientific findings and constitute a bridge to a theology of nature. In this way, philosophy, if correctly understood, can avoid the dangers of both using specific scientific results as the possible foundations of theology of nature and leaving those results as being uninterpreted factuality that loses any relation with the dimension of intelligibility. We have stressed that the science of the twentieth century has led to a new and fresh understanding of the dimension of chance and novelty in our universe, in a moment in which the scientific community was tempted to believe that an all-embracing scientific explanation of reality had been already accomplished. Therefore, to understand and assimilate the significance of this experience represents an epistemological warning against any lure to the closure of empirical knowledge. There are signs that we are again approaching a similar situation in current scientific developments. The philosophical examination of this point has brought us to the understanding that natural laws cannot be considered as ruling single happenings. We have also stressed that such an awareness does not imply any fundamental limitation in scientific knowledge since science progresses precisely through stimuli coming from unexpected results, and has the duty to provide explanations that are more and more general and therefore less and less anchored to the details of physical reality. All these considerations constitute the contents of the first heuristic principle that we have assumed: happenings are ontologically irreducible to any regularity. We have further asked about the ground of the astonishing agreement between our theories and nature. We have pointed out that if e.g. quantum systems were to show only a chance behaviour, we would be unable to formulate those predictions that characterize quantum mechanics and even represent its beauty. The ontological root of this ability to predict is represented by correlations, that is, interdependencies among physical systems that confine the space of the possible happenings to a subset of the ensemble of all random events. This is the source of order in our world and therefore the ground of its intelligibility. The reduction of the space of the possible events is strictly connected with the issue of information. Indeed, we interpret those correlations as formal constraints showing the fact that nature is not only characterized by physical parameters but also by informational ones. This is the content of the second heuristic principle that we have introduced. Now, the critical point is that we would have a major conflict between these two principles if the nature of those correlations were not potential and general, thus allowing for the arising of chance happenings. Moreover, we have stressed that there is a dynamic interplay between happenings and correlations, which represents the content of the third heuristic principle. The field in which this dynamic interplay was first considered and where it is displayed with major evidence is biology. However, we have also shown that this behaviour is ubiquitous in nature. Such an interplay means that a system, when perturbed, is driven to find new states of transient equilibrium by integrating both internal and external constraints. In this way, we can overcome the traditional science of fixed attractors and focus on the fundamental concepts of dynamic itinerancy and evolving attractors. We have in particular shown that such an understanding of nature allows to grasp the continuous emergence of new variety, new regularities and therefore also the growing levels of complexity. Indeed, when a system is perturbed it may activate new constraints (that were previously dormant) and this in turn will canalize the further process and especially the search for new equilibria. So, we have shown that necessity arises in our universe as a consequence of dynamism and not as a presupposition which determines that dynamism. Since this behaviour cuts across all domains of nature, we have spoken of emergent monism, though not in a traditional materialist sense (as we have said, formal aspects are also relevant at any level of complexity) but rather in the sense that nature shows an incredible simplicity in its complexity. We have also pointed out that, if the tendency to disorder is spontaneous, it is likely that there is also a compensating tendency to order, since according to quantum mechanics the universe should preserve the same amount of order and disorder as a whole. This could explain why the universe and in particular biological evolution show an amazing tendency to build new and higher levels of order.

Such a philosophical view allows us to understand the world theologically as being ruled by the divine Providence and brought to existence through God's continual act of creation, without implying any form of pre-constituted design on the model of an anthropomorphic or Demiurge-like deity. The distinction between the primary Cause and the secondary causes allows us to consider God as the direct cause of any existence whilst the finite causes are directly responsible for the specific effects that they induce on other beings. Moreover, this state of affairs also enables us to participate in the arising of new beings through the activation of formal constraints. This is an issue that is very relevant across the history of evolution on our planet. We have also clarified that we do not need to consider the contingent features that characterize humanity as a biological species on Earth as targeted in advance. On the contrary, we have stressed that humans in their universal value as beings endowed with rationality, using symbols, displaying intentionality, and finally grasping God as the ultimate ground of any intelligibility and power of our universe, have a privileged place in creation, since through their knowledge and love of the universe, they allow for its elevation to a spiritual communion with its Creator. The specific way through which humans do this is self-canalization, that is, the capability of human beings to freely accept, as a consequence of their own enquiry and search for truth, principles (even moral principles), rules and laws that ultimately do not depend on their arbitrary will, to assimilate them and to transmit their results to the subsequent generations, grounding what we call culture. We have stressed that a deep and ripe understanding of humanity, and therefore a suitable integration of naturalism and anthropology, should be accomplished through a critical merging of the three major philosophical approaches of the Western tradition: Platonism, whose focus is the world of intelligibility and formal constructs; Aristotelism, centred on the fundamental dynamicity of our physical world; modern Idealism as focussing on the spontaneous activity of the human mind and its ability to grasp truths of universal value.

A fundamental methodological question is the following: if in the course of the cognitive journey of humanity we try to reach conclusions that are more and more general, but are forced to correct our previous assumptions when we have received surprising inputs or data, then what can the cognitive value of philosophy and still more of theology be since they presume to deal with wholes that are much ahead of any scientific research program (and therefore of current empirical testability)? Any knowledge is aspiration to know the whole, to know what philosophers call *being*. Therefore, it is a necessity of humanity to heuristically anticipate what the general direction of the cognitive journey could be, that ideally closes the circle, so to say. Theology, critically assisted by philosophy, is able to anticipate wholes.

This anticipation can back-affect the way in which we do our research. It can wrongly back-affect the research, as it happened with the Galilei affair, but it can also improve scientific research and give further support to certain strategies and therefore enhance the search for truth. Heuristics and thinking about first principles is not less important than dealing with facts. Obvioulsy, the progress of knowledge will feedback on philosophy and theology as well, so that a positive circulation of ideas could be established. This does not mean that the relations between these three fields have always been or should always be idyllic. As a matter of fact, as documented in this book, the Modern Ages until recent times have been dominated by a sense of fracture and contraposition among the different fields that are the object of this study. As a conclusion, we would like to point out that this modern fracture among science, philosophy, and theology does not only have a negative value as many, even in the Catholic Church, still assume. We think that, due to a long-lasting and sharp process of confrontation, all these three fields are today much riper than at the beginning of the Modern Ages. In particular, they have learnt to recognize the methodology and the contributions of the other two fields, at least to a certain extent. For this reason, they are much more aware of their own specific methods, competences, and goals. We think about the theological developments occurred in biblical hermeneutics, the discussions on the foundations of quantum theory developed by giants like Einstein and Bohr, and the incredible contribution to epistemology and theory of knowledge provided by twentieth century philosophy. So, we may say that the mentioned fracture has been beneficial to a certain extent. If we are right in evaluating that the situation in the current developments in natural sciences is changing deeply, then the context could become ripe for a new assessment and hopefully for an integration of these three fields, and we hope that our three heuristic principles could be considered as a new and updated way to deal with this integration, a way that certainly was less practicable in the past centuries. When we say "integration" we mean a cognitive framework in which each of these fields preserves not only its full autonomy but also its essential difference relative to the other two. In this sense the situation today is very different relative to both the beginning of the Modern Ages but also to the Middle Ages, where a strong continuity between those fields was assumed. The future will say whether or not this conjecture is right.

APPENDIX

If the potential energy expressing the action of an external field on a mechanical system is a homogeneous function of the coordinates x_1, x_2, ..., x_n, that is, if it is a function satisfying (in the one-dimensional case for the sake of simplicity):

$$U(ax_1, ax_2, ..., ax_n) = a^k U(x_1, x_2, ..., x_n),$$

where a is an arbitrary constant, and, given the following transformations of space and time

$$x_m \rightarrow ax_m, \quad t \rightarrow bt,$$

we have that all velocities $v_m = dx_m/dt$ will vary a/b times whilst the kinetic energy $mv^2/2$ varies a^2/b^2 times. If a and b are ruled by the following equation

$$a^2/b^2 = a^k,$$

which implies

$$b = a^{(1 - k/2)},$$

all equations of motion remain the same after multiplication with the constants a and b. In other words, using the constants a and b means to transform all trajectories of the involved systems in other trajectories that are geometrically similar but different only in the linear (space and time) dimensions (due a shift in space and time). Therefore, different times t and t' of motion (taken on two points of the trajectories) will be in the relation

$$t'/t = (l'/l)^{(1 - k/2)},$$

where l'/l is the relation between the corresponding unidimensional spatial dimensions. Given these assumptions, speed, energy and angular momentum will be in similar relations. Now, it is interesting to investigate what happens for different values of the parameter k. If we take $k = 1$, we obtain

$$t'/t = (l'/l)^{1/2},$$

which is Galilei's law of the fall of bodies on the Earth (the square of the times of fall are proportional to the heights of fall). If we take $k = -1$, we obtain

$$t'/t = (l'/l)^{3/2},$$

which is the third law of Kepler, stating that the squares of the times of the revolution of planets are proportional to the cubes of the orbit dimension. As is well known, from this law, we can derive Newton's gravitational law, and this was also the historical path followed by Newton himself[1]. Therefore, it is fully possible to derive the gravitational law from general mechanical considerations about the potential energy and the possible trajectories followed by the involved systems.

---

[1] Koyré 1968.

REFERENCES

1. Catholic Church's Documents

Aeterni Patris Leo XIII, *Aeterni Patris. Encyclical of Pope Leo XIII on the Restoration of Christian Philosophy*, given at St Peter's, Rome August the 4th, 1879.

Caritas in Veritate Benedict XVI, *Encyclical Letter Caritas in Veritate of the Supreme Pontiff Benedict XVI to the Bishops, priests and deacons men and women religious the lay faithful and all people of good will on integral human development in charity and truth*, given at Saint Peter's, Rome, on June the 29th, 2009.

Fides et Ratio John Paul II, *Encyclical letter Fides et Ratio of the Supreme Pontiff John Paul II to the Bishops of the Catholic Church on the Relationship between Faith and Reason*, given at Saint Peter's, Rome, September the 14th, 1998.

Gaudium et Spes Paul VI, *Gaudium et Spes*, Pastoral Constitution of the Second Vatican Council, promulgated by Pope Paul VI on December 7, 1965.

Providentissimus Deus Leo XIII, *Providentissimus Deus. Encyclical of Pope Leo XIII on the study of the Holy Scriptures*, given at St. Peter's, Rome, November the 18th, 1893.

2. Authors' References

Alberts *et al.* 2008 Alberts B., Johnson A., Lewis J., Raff M., Roberts K., Walter P., *The Molecular Biology of the Cell*, New York 2002, 2008.

Alexander 2001 Alexander D., *Rebuilding the Matrix: Science and Faith in the 21st Century*, Oxford 2001.

Alexander 2008 Alexander D., *Creation or Evolution: Do We Have to Choose?*, Oxford (UK)-Michigan (USA) 2008.

Alexander–White 2004 Alexander R., White R.S., *Beyond Belief: Science, Faith and Ethical Challenges*, Oxford 2004.

Anagnostopoulos 2009a Anagnostopoulos G. (Ed.), *A Companion to Aristotle*, Malden (MA), 2009.

Anagnostopoulos 2009b — Anagnostopoulos G., "Aristotle's Methods", in Anagnostopoulos 2009a, pp. 101—122.

Aquinas *Con. Gen.* — Thomas Aquinas, *Summa Contra Gentiles*.

Aquinas *De Pot.* — Thomas Aquinas, *Quaestiones Disputatae de Potentia*.

Aquinas *De Ver.* — Thomas Aquinas, *Quaestiones Disputatae de Veritate*, Torino, 1964.

Aquinas *In Met.* — Thomas Aquinas, *In Arist. Metaphysyca*.

Aquinas *S. Th.* — Thomas Aquinas, *Summa Theologiae*, Torino, 1952.

Aquinas *Sup. Sent.* — Thomas Aquinas, *Super Sententiarum*.

Arber *et al.* 2011 — Arber W., Mittelstrass J., Sorondo M.M. (Eds.), *The Scientific Legacy of the 20th Century*, Vatican City 2011.

Aristotle *An. Pr.-An. Post.* — Aristotle, *Analytica Priora et Posteriora* (Eds. W.D. Ross and L. Minio-Paluello), Oxford 1964, 1986.

Aristotle *Cat.-Int* — Aristotle, *Categoriæ et Liber de Interpretatione*, Oxford 1949, 1986.

Aristotle *De Caelo* — Aristotle *De Caelo*, in Aristotle *Op.*, pp. 268-313.

Aristotle *Fisica* — Aristotle, *Fisica* (Ed. L. Ruggiu), Milano 2007.

Aristotle *Gen. An.* — Aristotle, *De generatione animalium*, in Aristotle *Op.*, pp. 715-789.

Aristotle *Gen. Corr.* — Aristotle, *De generatione et corruptione*, in Aristotle *Op.*, pp. 314-338.

Aristotle *Generation* — Aristotle, *De generatione et corruptione* (Ed. C.J.F. Williams), Oxford, 1982.

Aristotle *Met.* — Aristotle, *Metaphysica*, Oxford 1957.

Aristotle *Op.* — *Aristotelis Opera* (Ed. I. Bekker), Berlin, 1831-1882.

Aristotle *Part. An.* — Aristotle, *De Partibus Animalium*, in Aristotle *Op.*, pp. 639-697.

Aristotle *Phys.* — Aristotle, *Physica*, Oxford 1950, 1988.

Aristotle *Posterior An.* — Aristotle, *Posterior Analytics* (Ed. J. Barnes), Oxford, 1975.

Aristotle *Top.* — Aristotle, *Topica*, in Aristotle *Op.*, pp. 100-164.

Arnold 1978 — Arnold V.I., *Mathematical Methods of Classical Mechanics*, New York 1978; 2nd ed. 1989.

Augustine 399-422 — Augustine of Hyppo, *De Trinitate*.

Auletta 2000 — Auletta G., *Foundations and Interpretation of Quantum Mechanics: In the Light of a Critical-Historical Analysis of the Problems and of a Synthesis of the Results*, Singapore 2000; rev. ed. 2001.

Auletta 2002 — Auletta G., "Is Representation Characterized by Intrinsicity and Causality?", *Intellectica* **35**, 83-113.

Auletta 2003 — Auletta G., "Some Lessons of Quantum Mechanics for Cognitive Science", *Intellectica* **36-37**, 293-317.

Auletta 2005 — Auletta G., "Quantum Information as a General Paradigm", *Foundations of Physics* **35**, 787-815.

Auletta 2006a — Auletta G., "The Ontology Suggested by Quantum Mechanics", in Valore P. (Ed.), *Topics on General and Formal Ontology*, Monza 2006, 161-79.

Auletta 2006b	Auletta G., "The Problem of Information", in Auletta G. (Ed.), *The Controversial Relationships Between Science and Philosophy: a Critical Assessment*, Vatican City 2006, 109-127.
Auletta 2006c	Auletta G., "Einstein, Bohr e Schrödinger: Un esempio di dibattito filosofico-scientifico", *21mo Secolo. Scienza e Tecnologia* **3**, 43-46.
Auletta 2007a	Auletta G., "Science, Philosophy, and Religion Today: Some Reflections", *Theology and Science* **5**, 267-87.
Auletta 2007b	Auletta G., "Il dibattito tra Einstein, Bohr e Schrödinger: Causalità, determinazione e azione a distanza", *Il Protagora* **9**, 181-198.
Auletta 2007c	Auletta G., "Quale realtà si può attribuire agli universali?", *Chora* **14**, 15-20.
Auletta 2008	Auletta G., "How Many Causes Are There?", *21mo secolo. Scienza e tecnologia* **5**, 41-48.
Auletta 2009a	Auletta G., "What About the Three Forms of Inference?", *Acta Philosophica* **18**, 59-74.
Auletta 2009b	Auletta G., "Perchè la scienza è importante per la filosofia?", *Gregorianum* **90**, 335-354.
Auletta 2010	Auletta G., "A Paradigm Shift in Biology?", *Information* **1**, 28-59.
Auletta 2011a	Auletta G., *Cognitive Biology: Dealing with Information From Bacteria to Minds*, Oxford 2011.
Auletta 2011b	Auletta G., "Teleonomy: The feedback circuit involving information and thermodynamic processes", *Journal of Modern Physics* **2**, 136-45.
Auletta 2011c	Auletta G., "Correlations and Hyper-Correlations" *Journal of Modern Physics* **2**, 958-961.
Auletta 2011d	Auletta G., "Aristotele's Syllogistic and Natural Inferences of Scientific Kind" draft submitted for pubblication.
Auletta *et al.* 2008	Auletta G., Ellis G.F.R., Jaeger L., "Top-down causation by information control: from a philosophical problem to a scientific research programme", *The Journal of the Royal Society Interface* **5**, 1159-1172.
Auletta *et al.* 2009a	Auletta G., Fortunato M., Parisi G., *Quantum Mechanics*, Cambridge 2009.
Auletta *et al.* 2009b	Auletta G., Alai M., Tarozzi G., "Einstein's Local Realism and the Realistic Interpretation of the Wave Function", in Alunni C., Castellana M., Ria D., Rossi A. (Eds.), *Albert Einstein et Herman Weyl (1955-2005). Questions épistémologiques ouverte*, Taranto 2009, pp. 33-49.
Auletta *et al.* 2011a	Auletta G., Colagè I., D'Ambrosio P., "The game of life implies both teleonomy and teleology", in press.
Auletta *et al.* 2011b	Auletta G., Colagè I., D'Ambrosio P., "A critical assessment of transhumanism", in press.

Auletta *et al.* 2011c — Auletta G., Leclerc M, and Martinez R. (Eds.), *Biological Evolution: Facts and Theories: A Critical Appraisal after 150 years after "The Origin of Species"*, Rome 2011.

Auletta-Stoeger 2010 — Auletta G., Stoeger W., "Highlights of the Pontifical Gregorian University's International Conference on Biological Evolution", *Theology and Science* **8**, 7-15.

Auletta–Tarozzi 2004a — Auletta G., Tarozzi G., "Wave-like Correlations versus Path Detection: Another Form of Complementarity, *Foundations of Physics Letters* **17**, 889-895.

Auletta–Tarozzi 2004b — Auletta G., Tarozzi G., "On the Reality of Quantum Waves", *Foundation of Physics* **34**, 1675-1694.

Auletta–Torcal 2011 — Auletta G., Torcal L., "From Wave-Particle to Features-Event Complementarity", *International Journal of Theoretical Physics*, DOI 10.1007/s10773-011-0833-8.

Averroes AM — Averroes, *On Aristotle's "Metaphysics"*, (Tr. and Ed. R. Arnzen), 2009.

Averroes *PTA* — Averroes, *The Philosophy and Theology of Averroes* (Ed. M. Jamil-Ur-Behman), Baroda, 1921.

Averroes *DD* — Averroes, *Decisive Discourse on the Delineation of the Relation between Religion and Philosophy*, in Averroes *PTA*.

Avicenna *APP* — Avicenna, *Avicene perhypatetici philosophi* (Ed. Caecilius Fabrianesis), 1508; rep. Frankfurt a. M., 1961.

Avicenna *de An.* — Avicenna, *De anima*, in Avicenna *APP*.

Avicenna *GR* — Avicenna, *Livre de la genèse et du retour* (Ed. and tr. Y. J. Michot), Oxford 2002.

Ayala 1998a — Ayala F.J., "Darwin's Devolution: Design Without Designer", in Russell *et al.* 1998: 1-15.

Ayala 1998b — Ayala F.J., "Teleological Explanations versus Teleology", *History and Philosophy of the Life Sciences* **20**, 41-50.

Ayala 2006 — Ayala F.J., *Darwin and Intelligent Design*, Minneapolis 2006.

Ayala 2007 — Ayala F.J., *Darwin's Gift to Science and Religion*, Washington 2007.

Bacon 1620 — Bacon F., *Novum Organum*, London 1620; Hamburg 1990.

Barbour 1966 — Barbour I., *Issues in Science and Religion*, Englewood Cliffs 1966.

Barrow 1990 — Barrow J.D., *Theories of Everything*, Oxford 1990, 1991.

Barrow 2005 — Barrow J.D., *The Infinite Book: A Short Guide to the Boundless, Timeless and Endless*, New York 2005.

Barrow–Tipler 1986 — Barrow J.D., Tipler F.J., *The Anthropic Cosmological Principle*, Oxford 1986, 1987.

Barthes 1957 — Barthes R., *Mythologies*, Paris 1986.

Batterman 2002 — Batterman R.W., *The Devil in the Details: Asymptotic Reasoning in Explanation, Reduction, and Emergence*, Oxford 2002.

Baudrillard 1981 — Baudrillard J., *Simulacres et simulation*, Paris 1981.

Behe 1996   Behe M.J., *Darwin's Black Box*, 1996; New York 1998.

Bell 1964   Bell J.S., "On Einstein Podolsky Rosen paradox", *Physics* **1**, 195-200 (Repr. in Bell 1987, pp. 14-21).

Bell 1966   Bell J.S., "On the problem of hidden variables in quantum mechanics", *Review of Modern Physics* **38**, 447-452 (Repr. in Bell 1987, pp. 1-13).

Bell 1987   Bell J.S., *Speakable and Unspeakable in Quantum Mechanics*, Cambridge 1987.

Benedict XVI 2011   Benedict XVI, "Address", in Arber *et al.* 2011, pp. 23-24.

Berkeley 1709   Berkeley G., *An Essay Towards a New Theory of Vision*, rep. in G. Berkeley, *Works on Vision*, 1963; Westport, Conn., 1981, pp. 19-97.

Berkeley 1710   Berkeley G., *A Treatise Concerning the Principles of Human Knowledge*, Dublin, 1710, London, 1734; 1988.

Bhaskar 1975   Bhaskar R., *A Realist Theory of Science*, 1975; London 2008.

Blackmore 1999   Blackmore S., *The Meme Machine*, Oxford 1956.

Blair 1992   Blair G.A, *Energeia and Entelecheia: "Act" in Aristotle*, Ottawa 1992.

Bochenski 1956   Bochenski I. N., *Formale Logik*, Freiburg 1956.

Bohm 1952   Bohm D., "A Suggested Interpretation of the Quantum Theory in Terms of 'Hidden Variables'", *Physical Review* **85**, 166-193.

Bohm 1953   Bohm D., "Proof that Probability Density Approaches $\psi^*\psi$ in Causal Interpretation of the Quantum Theory", *Physical Review* **89**, 458-466.

Bohm 1980   Bohm D., *Wholeness and Implicate Order*, London 1980.

Bohm-Hiley 1987   Bohm D. and Hiley B. J., "An Ontological Basis for the Quantum Theory I: Non Relativistic Particle Systems", *Physics Reports* **144**, 323-348.

Bohm et al. 1987   Bohm D., Hiley B. J., and Kaloyerou P. N., "An Ontological Basis for the Quantum Theory II: A Causal Interpretation of Quantum Fields", *Physics Reports* **144**, 349-375.

Bohm-Hiley 1989   Bohm D. and Hiley B. J., "Non-Locality and Locality in the Stochastic Interpretation of Quantum Mechanics", *Physics Reports* **172**, 93-122.

Bohm-Hiley 1993   Bohm D. and Hiley B. J., *The Undivided Universe. An Ontological Interpretation of Quantum Theory*, London 1993.

Bohr 1913   Bohr N., "On the constitution of atoms and molecules", *Philosophical Magazine* **26**, 1-25, 476-502, 857-875.

Bohr 1920   Bohr N., "Über die Linienspektren der Elemente", *Zeitschrift für Physik* **2**, 423-469.

Bohr 1928   Bohr N., "The quantum postulate and the recent development of atomic theory", *Nature* **121**, 580-90.

Boltzman 1905   Boltzmann L., *Populäre Schriften*, Leipzig 1905.

Bondi 1948   Bondi H., "Review of Cosmology", *Monthly Notices of the Royal Astronomical Society* **108**, 104-120.

Bondi 1952   Bondi H., "Relativity and Indeterminacy", *Nature* **169**, 660.

Bondi-Gold 1948   Bondi H., Gold T., "The Steady-State Theory of the Expanding Universe", *Monthly Notices of the Royal Astronomical Society* **108**, 252-270.

Boscovich 1758   Boscovich R.J., *Theoria philosophiae naturalis redacta ad unicam legem virium in natura existentium*, Vienna, 1758; Venezia, 1763.

Bostrom-Sandberg 2008   Bostrom N., Sandberg A. "The Wisdom of Nature: An Evolutionary Heuristic for Human Enhancement" in Savulescu J., Bostrom N. (eds.), *Human Enhancement*, Oxford 2008, pp. 375-416.

Brune *et al.* 1996 Brune M., Hagley E., Dreyer J., Maître X., Maali A., Wunderlich C., Raimond J.M., Haroche S., "Observing the progressive decoherence of the 'meter' in a quantum measurement", *Physical Review Letters* **77**, 4887-4890.

Bultmann 1941   Bultmann R, "Neues Testament und Mythologie. Das Problem der Entmythologisierung der neutestamentlichen Verkündigung", in: Bartsch H.W. (Ed.), *Kerygma und Mythos*, Hamburg 1960: v. 1, pp. 15-48.

Cajetan 1514   Cajetan, Th. de Vio, *In Primam Sanctissimi doctoris Thomae Aquinatis Summae Theologiae partem*, Paris, 1514.

Cajetan 1568   Cajetan, Th. de Vio, *Prima Secundae Partis Summae Sacrae Theologiae Sancti Thomae Aquinatis*, Paris 1519; Antwerpen, 1568.

Carnap 1928   Carnap R., *Der Logische Aufbau der Welt*, Hamburg 1928.

Carnap 1932   Carnap R., "Die physikalische Sprache als Universalsprache der Wissenschaft", *Erkenntnis* **1**, 432.

Carnap 1936   Carnap R., "Testability and Meaning" *Philosophy of Science* **3**, 419-471.

Cartwright 1999   Cartwright N., *The Dappled World: A Study of the Boundaries of Science*, Cambridge 1999.

Cassirer 1910   Cassirer E., *Substanzbegriff und Funktionsbegriff*, Berlin, 1910; Darmstadt 1994.

Chaisson 2001   Chaisson E.J., *Cosmic Evolution: The Rise of Complexity in Nature*, Cambridge (MA) 2001.

Changeux 1997   Changeux J.-P., *Neuronal Man: The Biology of Mind*, Princeton 1997.

Changeux 2009   Changeux J.-P., *The Physiology of Truth: Neuroscience and Human Knowledge*, Cambridge (MA) and London 2009.

Changeux–Danchin 1976   Changeux J.-P., Danchin A., "Selective stabilization of developing synapses as a mechanism for the specification of neural networks", *Nature* **264**, 705-712.

Chomsky 1959   Chomsky N., "A Review of B. F. Skinner's Verbal Behavior", *Language* **35**, 26-58.

Chomsky 1992   Chomsky N., "Explaining Language Use", *Philosophical Topics* **20**, 205-231.

Chomsky 1995a   Chomsky N., *The Minimalist Program*, Cambridge (MA) 1995, 2001.

Chomsky 1995b   Chomsky N., "Language and Nature", *Mind* **104**, 1-61.

Clagett 1955   Clagett M., *Greek Science in Antiquity*, New York, 1955; Mineola, 2001.

Clarke 2010   Clarke P. G. H., "Determinism, Brain Function and Free Will", *Science & Christian Belief* **22**, 133-149.

Clayton 2004   Clayton P., *Mind and Emergence: From Quantum to Consciousness*, Oxford 2004.

Clayton–Davies 2006   Clayton P., Davies P.C.W. (Ed.), *The Re-Emergence of Emergence*, Oxford 2006.

Colagè 2006   Colagè I., "Between Realism and Instrumentalism: Description, Interaction and Truth. A Proposal", in Auletta G. (Ed), *The Controversial Relationships Between Science and Philosophy: a Critical Assessment* Vatican City 2008, pp. 303-315.

Colagè 2008   Colagè I., "Experiments and Causality", in Auletta G. (Ed), *The Relationships Between Science and Philosophy: New Opportunities for a Fruitful Dialogue*, Vatican City 2008, pp. 157-179.

Colagè 2010   Colagè I., *Interazione ed inferenza: Epistemologia scientifica ispirata al pensiero di C. S. Peirce*, Roma, 2010.

Colagè–Fano 2008   Colagè I., Fano G., "Teologia e Scienza: quale dialogo?", *Hermeneutica*, Nuova Serie 2008, 301-323.

Compton 1923   Compton A.H., "A quantum theory of the scattering of x-rays by light elements", *Physical Review* **21**, 483-502.

Comte 1830-1842   Comte A., *Cours de philosophie positive*, Paris 1830-1842.

Conway Morris 2006   Conway Morris S., *Life's Solution: Inevitable Humans in a Lonely Universe*, Cambridge 2006.

Conway Morris 2008   Conway Morris S., "Evolution and Convergence: Some Wider Considerations", in Conway Morris S. (Ed.), *The Deep Structure of Biology*, West Conshohocken (PA), 2008.

Cottier 2011   Cottier G., "The Metaphysical Meaning of Creation" in Auletta *et al.* 2011c, pp. 419-32.

Coyne 1998 — Coyne G.V., "Evolution and the Human Person: The Pope in Dialogue", in Russell *et al.* 1998, pp. 11-17.

Coyne 2005 — Coyne G.V., "Destiny of Life and Religious Attitudes", in Seckbach J. (Ed), *Life as We Know It*, Dortrecht 2005, pp. 521-535.

Coyne 2007 — Coyne G.V., *Faith and Knowledge: Towards a New Meeting of Science and Theology*, Vatican City 2007.

Crick 1958 — Crick F., "On Protein Synthesis", *Symp. Soc. Exp. Biol.* **12**, 138-63.

Crick 1970 — Crick F., "Central Dogma of Molecular Biology", *Nature* **227**, 561-563.

Crutchfield *et al.* 1986 — Crutchfield J.P., Farmer J.D., Packard N.H., Shaw R.S., "Chaos", *Scientific American* (dec. 1986): 46-57.

Cudworth 1678 — Cudworth R. D. D., *The True Intellectual System of the Universe, wherein all the Reason and Philosophy of Atheism is Confuted, and its Impossibility Demonstrated*, London 1678.

Dalton 1808 — Dalton J., *A New System of Chemical Philosophy*, Manchester 1808.

D'Ambrosio 2008 — D'Ambrosio P., "Analysis without Reductionism? Ernst Mayr and the Uniqueness of Biology", in Auletta G. (Ed) *The Relationships Between Science and Philosophy: New Opportunities for a Fruitful Dialogue*, Vatican City 2008, pp. 181-186.

Damasio 1999 — Damasio A., *The Feeling of What Happens: Body and Emotion in the Making of Consciousness*, New York 1999.

Darwin 1859 — Darwin C.R., *The Origin of Species*, London 1859; New York 1999.

Darwin *Corr.* — *The Correspondence of Charles Darwin*, F. Burkhardt, *et al.* (Eds), Cambridge (UK) 1985-.
Also at: www.darwinproject.ac.uk/the-correspondence.

Dawkins 1976 — Dawkins R., *The Selfish Gene*, Oxford 1976, 1989, 1999.

Dawkins 1983 — Dawkins R., "Universal Darwinism", in Bendall D.S. (Ed), *Evolution from Molecules to Man*, Cambridge 1983, pp. 403-425.

Dawkins 1986 — Dawkins R., *The Blind Watchmaker*, Harlow 1986.

Dawkins 2006 — Dawkins, R., *The God Delusion*, Bantam Press, 2006; Black Swan, 2007.

De Beer 1971 — De Beer G.R., *Homology: An Unsolved Problem*, Oxford 1971.

Dembski 1998 — Dembski W.A., *Design Inference: Eliminating Chance Through Small Probabilities*, Cambridge 1998; 2005.

Dembski 1999 — Dembski W.A., *Intelligent Design: The Bridge Between Science and Theology*, Downers Grove 1999.

De Morgan 1847 — De Morgan A., *Formal Logic or the Calculus of Inferences, Necessary and Probable*, London 1847.

Dennett 1978 — Dennett D., *Brainstorms*, Montgomery (VT) and Bradford 1981, 1985.

Dennett 1987   Dennett D., *The Intentional Stance*, Cambridge (MA), 1987, 1989, 1998.

Dennett 1995   Dennett D., *Darwin's Dangerous Idea*, New York 1995.

Derrida 1967   Derrida J., *L'écriture et la différence*, Paris 1967.

Derrida 1972   Derrida J., *La Dissémination*, Paris 1972.

Descartes 1637   Descartes R., *Discours de la méthode pour bien conduire sa raison, et chercher la vérité dans les sciences*, Leiden 1637.

Descartes 1641   Descartes R., *Meditationes de prima philosophia*, Paris 1641.

Descartes 1644   Descartes R., *Principia Philosophiae*, Amsterdam 1644.

Draper 1875   Draper J. W., *History of the Conflict between Religion and Science*, 1875.

Duhem 1906   Duhem P., *La théorie physique. Son object, sa structure*, Paris 1906.

Duhem 1913-1959   Duhem P., *Le système du monde*, Paris 1913-1959.

Duns Scot *Ord.*   Duns Scot J., *Ordinatio*, in Duns Scot J., *Opera Omnia*, Vatican City 1950-.

Duns Scot *Rep. Par.*   Duns Scot J., *Reportata Parisiensa*, in Duns Scot *OO*, vv. 22-24.

Duns Scot *OO*   Duns Scot J., *Opera Omnia*, Editio Nova, Paris 1991-1995.

Dyson 2008   Dyson F., *The Scientist as a Rebel*, New York 2008.

Eccles 1970   Eccles J.C., *Facing Reality*, Berlin 1970.

Edelman 1987   Edelman G.M., *Neuronal Darwinism: The Theory of Neuronal Group Selection*, New York 1987.

Edelman 1992   Edelman G.M., *Bright Air, Brilliant Fire. On the Matter of Mind*, New York 1992.

Edelman–Tononi 2000   Edelman G.M., Tononi G., *A Universe of Consciousness*, New York 2000.

Edwards 1999   Edwards D., *The God of Evolution: A Trinitarian Theology*, New York 1999.

Edwards 2010   Edwards D., *How God Acts: Creation, Redemption and Special Divine Action*, Minneapolis (MN) 2010.

Einstein 1905   Einstein A., "Über einen die Erzeugung und Verwandlung des Lichtes betreffenden heuristischen Gesichtspunkt", *Annalen der Physik* **17**, 132-148.

Einstein 1918   Einstein A., "Prinzipien der Forschung", Repr. in Einstein *MW*, pp. 107-113.

Einstein 1926   Einstein A., "Letter to Max Born", in *The Born-Einstein Letters*, New York 1971.

Einstein 1936   Einstein A., "Physik und Realität", *Journal of Franklin Institute* **221**, 349-382 (Eng. tr. in Einstein 1956, pp. 59-97).

Einstein 1937   Einstein A., "Moral Decay", in Einstein 1956, pp. 7-8.

Einstein 1939   Eistein A., "Science and Religion" in Einstein 1956, pp. 19-28.

Einstein 1942	Eistein A., "Isaac Newton" in Einstein 1956, pp. 219-223.
Einstein 1956	Einstein A., *Out of My Later Years: Estate of A. Einstein*, New York 1956, 1993.
Einstein *MW*	Einstein A., *Mein Weltbild*, Frankfurt 1993.
Einstein *et al.* 1935	Einstein A., Podolsky B., Rosen N., "Can quantum-mechanical description of physical reality be considered complete?", *Physical Review* **47**, 777-780.
Einstein–Infeld 1938	Einstein A., Infeld L., *The Evolution of Physics*, New York 1938.
Eliade 1958	Eliade M., *Patterns in Comparative Religion*, New York 1958.
Ellis 2004	Ellis G.F.R., "Physics, Complexity and Causality", *Nature* **435**, 743.
Ellis 2005	Ellis G.F.R., "Physics and the Real World", *Foundations of Physics* **35**, 49.
Fabro 1941a	Fabro C., *La fenomenologia della percezione*, Milano; 2nd ed. Brescia, 1961; Repr. in Fabro *OC*, v. 5.
Fabro 1941b	Fabro C., *Percezione e pensiero*, Milano, 1941; 2nd ed. Brescia 1962; Repr. in Fabro *OC*, v. 6.
Fabro *OC*	Fabro C., *Opere Complete*, Segni (Rome), 2005-.
Farber 2005	Farber I., "Peirce on Reality, Truth, and the Convergence of Inquiry in the Limit", *Transactions of the Charles S. Peirce Society* **41**, 541-566.
Ferejohn 2009	Ferejohn M., "Empiricism and the First Principles of Aristotelian Science", in Anagnostopoulos 2009a, pp. 66-80.
Feyerabend 1969	Feyerabend P.K., "On a recent critique of complementarity. II", *Philosophy of Science* **36**, 82-105.
Feyerabend 1976	Feyerabend P.K., *Against Method. Outline of an Anarchistic Theory of Knowledge*, New Left Books 1976.
Feynman 1988	Feynman R.P., "String Theory", in: P.C.W. Davis and J. Brown (Eds), *Superstrings: a Theory of Everything?*, Cambridge 1998, pp. 192-210.
Fichte 1794	Fichte J.G., *Grundlage der gesamten Wissenschaftslehre*, Leipzig 1794, Hamburg 1988.
Fichte 1804	Fichte J.G., *Die Wissenschaftslehre. Zweiter Vortrag im Jahre 1804*, 1908, 1975, Hamburg 1986.
Fodor 1987	Fodor J., *Psychosemantics: The Problem of Meaning in the Philosophy of Mind*, Cambridge (MA) 1987, 1998.
Fodor and Piattelli-Palamarini 2010	Fodor J. A. and Paittelli-Palmarini, M., *What Darwin Got Wrong*, London 2010.
Foucault 1969	Foucault M., *L'archéologie du savoir*, Paris 1969.
Foucault 1972	Foucault M., *Histoire de la folie à l'âge classique*, Paris 1972.
Frank 1966	Frank R. M., "The Structure of Created Causality in Al-As'ari", *Stu-*

dia Islamica **25**, 2-75.

Frede–Patzig 2001 — Frede M, Patzig G., *Il Libro Z della Metafisica di Aristotele*, Milano 2001.

Frege 1884 — Frege G., *Die Grundlagen der Arithmetik: eine logisch-mathematische Untersuchung über den Begriff der Zahl*, Breslau, 1884.

Friedman *et al.* 2000 — Friedman J. R., Patel V., Chen W., Tolpygo S. K., and Lukens J. E., "Quantum Superposition of Distinct Macroscopic States", *Nature* **406**, 43-46.

Friston and Kiebel 2009 — Friston, K. and Kiebel, S., "Predictive Coding Under the Free-Energy Principle", *Philosophical Transactions of the Royal Society* **B364**: 1211-1121.

Friston and Stephan 2007 — Friston, K. J. and Stephan, K. E., "Free–Energy and the Brain", *Synthese* **159**, 417-458.

Friston *et al.* 2010 — Friston, K. J., Kilner, J., and Harrison, L., "A Free Energy Principle for the Brain", *Journal of Physiology* **100**, 70-87.

Futuyma 1998 — Futuyma D.J., *Evolutionary Biology*, Sunderland (MA) 1998.

Galilei 1612 — Galilei G., "Lettera a Marco Velseri", 1 dec. 1612.

Galilei 1615 — Galilei G., "Lettera a Cristina de Lorena", 1615, (Repr. in *Edizione nazionale delle Opere di Galileo Galilei*, 1968, vol. V, 307-348).

Galilei 1623 — Galilei G., *Il Saggiatore*, Roma 1623.

Galilei 1632 — Galilei G., *Dialogo sopra i due massimi sistemi del mondo*, Firenze 1632, Torino 1970.

Gentile 1913 — Gentile G., *La riforma della dialettica hegeliana*, Messina 1913.

Giberson–Artigas 2007 — Giberson K., Artigas M., *Oracles of Science: Celebrity Scientists versus God and Religion*, New York 2007.

Gilbert 1600 — Gilbert W., *De Magnete, Magneticisque Corporibus, et de Magno Magnete Tellure*, London, 1600; 2nd ed. Stettin, 1628; 3rd ed. 1633; rep. of the 1st ed. 1892; Eng tr. 1893.

Gilbert *et al.* 1996 — Gilbert S.G., Opitz J.M., Raff R.A., "Resynthesizing Evolutionary and Developmental Biology", *Developmental Biology* **173**, 357-372.

Gilbert 2008 — Gilbert S.F., *Developmental Biology*, Sunderland (MA) 2008.

Gilbert–Epel 2009 — Gilbert S.F., Epel D., *Ecological Developmental Biology: Integrating Epigenetics, Medicine, and Evolution*, Sunderland (MA) 2009.

Gilley 2009 — Gilley S., "Life and Writings", in Ker-Merrigan 2009, pp. 1-28.

Gilson 1944 — Gilson E., *La philosophie au moyen âge*, Paris 1944.

Gilson 1960 — Gilson E., *Introduction à la philosophie chrétienne*, Paris 1960.

Gilson 1987 — Gilson E., *L'être et l'essence*, Paris.

Gödel 1931 — Gödel K., "Über formal unentscheidbar Sätze der *Principia Mathematica* und verwandter Systeme", *Monatshefte für Mathematik und Physik* **38**: 173-198.

Gordon 1999 — Gordon R., *The Hierarchical Genome and Differentiation Waves*, Singapore 1999.

Gould 2002 — Gould S.J., *The Structure of Evolutionary Theory*, Cambridge (MA) 2002.

Gould–Lewontin 1979 — Gould S.J., Lewontin R., "The Sprandels of San Marco and the Panglossian Paradigm", *Proceedings of the Royal Society of London* **B205**, 581-598.

Gröblacher *et al.* 2007 — Gröblacher S., Paterek T., Kaltenbaek R., Brukner C, Zukowski M., Aspelmeyer M., and Zeilinger A., "An Experimental Test of Non-Local Realism", *Nature* **446**, 871-75.

Grünbaum 1973 — Grünbaum A., *Philosophical Problems of Space and Time*, Dordrecht 1973.

Guessoum 2011 — Guessoum N., *Islam's Quantum Question: Reconciling Muslim Tradition and Modern Science*, London-New York 2011.

Guillelmus a Sancto Theodorico, *Exp. Ro.* — Guillelmus a Sancto Theodorico, *Expositio super Epistolam ad Romano*.

Gupta–Lewontin 1982 — Gupta A.P., Lewontin R.C., "A Study of Reaction Norms in Natural Populations of Drosophila Pseudoobscura", *Evolution* **36**, 934-948.

Hall 1954 — Hall A.R., *The Revolution in Science. 1500-1750*, London 1954, 2nd ed. 1962, 1983, 1989.

Hamilton 1859 — Hamilton W., *Lectures on Metaphysics and Logic* (Eds. L. Mansel and J. Veitch), Edinburgh and London 1859.

Harré 1986 — Harré R., *Varieties of Realism: A Rationale for the Natural Sciences*, Oxford-New York 1986.

Harrison 2007 — Harrison P., *The Fall of Man and the Foundations of Science*, Cambridge, 2007.

Hartmann 1923-29 — Hartmann N., *Die Philosophie des Deutschen Idealismus*, Berlin 1923-29.

Hartmann 1950 — Hartmann N., *Philosophie der Natur Abriss der Speziellen Kategorienlehre*, Berlin 1950.

Hasnain *et al.* 1998 — Hasnain M., Fox P.T., Woldorff M.G., "Intersubject variability of functional areas in the human visual cortex", *Human Brain Mapping* **6**, 301-315.

Haughey 2009 — Haughey J.C., *Where is Knowing Going? The Horizon of the Knowing Subject*, Washington 2009.

Haught 2000 — Haught J. F., God *After Darwin: A Theology of Evolution*, 2000; Boulder (CO), 2001; 2nd ed. 2008.

Haught 2006 — Haught J F., *Is Nature Enough? Meaning and Truth in the Age of Science*, Cambridge 2006, 2007.

Haught 2010 — Haught J. F., *Making Sense of Evolution: Darwin, God and the Drama of Life*, Louisville (KY) 2010.

Hawking-Mlodinow 2010   Hawking, S. W., Mlodinow, L., *The Grand Design: New Answers to the Ultimate Questions of Life*, London 2010.

Heap–Comim 2008   Heap B., Comim F., "Consumption and Happiness: Christian Values and Sustainability", in Auletta G. (Ed), *The Relationships Between Science and Philosophy: New Opportunities for a Fruitful Dialogue*, Vatican City 2008, pp. 107-130.

Hegel 1807   Hegel G.W.F., *Phänomenologie des Geistes*, Bamberg–Würzburg 1807; Frankfurt 1970, 1973.

Hegel 1817   Hegel G.W.F., *Enzyklopädie der philosophischen Wissenschaft*, 1817; Frankfurt 1970, 1983.

Hegel 1821   Hegel G.W.F., *Grundlinien der Philosophie des Rechts oder Naturrecht und Staatswissenschaft im Grundrisse*, Berlin 1821, 1981.

Hegel 1833   Hegel G.W.F., *Wissenschaft der Logik*, 2nd ed. 1833; 1841; 1923–1932 (Ed. Lasson); Harmburg 1967.

Hegerfeldt 1974   Hegerfeldt G. C., "Remarks on Causality and Particle Localization", *Physical Review* **D10**, 3320-3321.

Hegerfledt 1985   Hegerfeldt G. C., "Violation of Causality in Relativistic Quantum Theory?", *Physical Review Letters* **54**, 2395-2398.

Heidegger 1928-29   Heidegger M., *Was ist Metaphysik?*, 1929; Frankfurt a. M., 4th ed. 1934; 6th ed. 1951.

Heidegger 1946   Heidegger M., "Brief über den "Humanismus", in Heidegger *GA*, v. 9, pp. 313-364.

Heidegger 1949   Heidegger M., "Einleitung zu "Was ist Metaphysik?"", in *GA*, v. 9, pp. 365-83.

Heidegger 1954   Heidegger M., *Vorträge und Aufsätze*, Pfullingen.

Heidegger 1955   Heidegger M., *Vom Wesen des Grundes*, 1955; Frankfurt a. M., 1995.

Heidegger *GA*   Heidgger M., *Gesamtausgabe*, Frankfurt, 1976 ff.

Heisenberg 1925   Heisenberg W., "Über quantentheoretische Umdeutung kinematischer und mechanischer Beziehungen", *Zeitschrift für Physik* **33**, 879-893.

Heisenberg 1927   Heisenberg W., "Über den anschaulichen Inhalt der quantentheoretischen Kinematik und Mechanik", *Zeitschrift für Physik* **43**, 172-198.

Heisenberg 1958   Heisenberg W., *Physics and Philosophy*, New York 1958.

Heller 2005   Heller M., *Some Mathematical Physics for Philosophers*, Vatican City 2005.

Heller 2008   Heller M., *Nuova fisica e nuova teologia*, Milano 2009.

Hempel 1953   Hempel C.G., *Methods of Concept Formation in Science*, Chicago 1953,

Hilbert 1903   Hilbert D., *Grundlagen der Geometrie*, Leipzig 1903.

Hilbert 1915 — Hilbert D., "Grundlagen der Physik", *Nach R. d. Kön. G. d. Wiss. zu Göttingen* 1915.

Hintikka 1964 — Hintikka J., "The Once and Future Sea Fight: Aristotle's Discussion on Future Contingents in De Int. IX'", *Philosophical Review* **73**, 461-492.

Hoffmann 1972 — Hoffmann B., *Albert Einstein: Creator and Rebell*, New York 1972.

Holder 2004 — Holder, R. D., *God, the Multiverse, and Everything: Modern Cosmology and the Argument from Design*, Aldershot 2004.

Horodecki *et al.* 2005 — Horodecki M., Oppenheim J., Winter A., "A partial quantum information", *Nature* **436**, 673-676.

Hoyle 1948 — Hoyle F., "A New Model for the Expanding Universe", *Monthly Notices of the Royal Astronomical Society* **108**, 372-382.

Hughes 2009 — Hughes G. J., "Conscience", in Ker-Merrigan 2009, pp. 189-220.

Hume 1748 — Hume D., *Enquiry Concerning Human Understanding*, 1748; 3rd ed. 1754.

Husserl 1950 — Husserl E., *Cartesianische Meditationen*, 1950; Hamburg 1977.

James 1878 — James W., "Remarks on Spencer's Definition of Mind as Correspondence", *Journal of Speculative Philosophy* **12**, 1-18.

James 1890 — James W., *The Principles of Psychology*, 1890; 1918; New York, 1950.

Jammer 1999 — Jammer M, *Einstein and Religion*, Princeton 1999.

Jeannerod 2006 — Jeannerod M., *Motor Cognition: What Actions Tell the Self*, Oxford 2006.

Jeannerod 2009 — Jeannerod M., *Le cerveau volontaire*, Paris 2009.

John Paul II 1981 — John Paul II, "Papal Address to the Pontifical Academy of Sciences 3 Oct. 1981", in *Papal Addresses to the Pontifical Academy of Sciences*, Vatican City 2003, pp. 249-52.

John Paul II 1988 — John Paul II, *Letter to Fr. Coyne*, in Russell *et al.* 1988.

John Paul II 1992 — John Paul II, "Papal Address to the Pontifical Academy of Sciences 31 Oct. 1992", in *Papal Addresses to the Pontifical Academy of Sciences*, Vatican City 2003, pp. 336-343.

Johnson 1991 — Johnson P.E., *Darwin on Trial*, Washington (DC) 1991.

Jou 2008 — Jou D., *Déu, cosmos, caos. Horitzons del diàleg entre ciència i religió*, Barcelona 2008.

Julsgaard *et al.* 2001 — Julsgaard B., Kozhekin A., and Polzik E. S., "Experimental Long-Lived Entanglement of Two Macroscopic Objects", *Nature* **413**, 400-403.

Kandel *et al.* 2000 — Kandel E.R., Schwartz J.H., Jessell T.M., *Principles of Neural Science*, 5th edition (International edition), New York 2000.

Kant 1787 — Kant I., *Kritik der Reinen Vernunft*, in *Kants Werke. Akademie Textausgabe*, Berlin 1968.

Katz 1954 — Katz J., "Plotinus and the Gnostics", *Journal of the History of Ideas* **15**, 289-298.

Kentros *et al.* 2004 — Kentros C.G., Agnihotri N.T., Streater S., Hawkins R.D., Kandel, E.R., "Increased Attention to Spatial Context Increases Both Place Field Stability and Spatial Memory", *Neuron* **42**, 283-295.

Kepler 1596 — Kepler J., *Mysterium cosmographicum*, 1596.

Kepler 1609 — Kepler J., *Astronomia nova*, 1609.

Kepler 1618-1621 — Kepler J., *Epitome astronomiae Copernicanae*, 3 volumes, 1618-1621.

Ker-Merrigan 2009 — Ker I. and Merrigan T. (Eds.), *The Cambridge Companion to John Henry Newman*, Cambridge 2009.

Kircher 1643 — Kircher A., *Magnes sive De arte magnetica Opus tripartitum*, Cologne 1643.

Kircher 1667 — Kircher A., *Magneticum Naturae Regnum sive Disceptatio physiologica*, Rome 1667.

Kircher 1680 — Kircher A., *Physiologia Kircheriana experimentalis*, Amsterdam 1680.

Koyré 1957 — Koyré A., *From the Closed World to the Infinite Universe*, Baltimore 1957.

Koyré 1966 — Koyré A., *Etudes galiléennes*, Paris 1966, 1980.

Koyré 1968 — Koyré A, *Etudes newtoniéennes*, Paris 1968.

Kuhn 1957 — Kuhn T.S., *The Copernican Revolution: Planetary Astronomy in the Development of Western Thought*, Cambridge (MA) 1957.

Kuhn 1961 — Kuhn T.S., "The Function of Measurement in Modern Physics", *Isis* **52**, 161-190.

Kuhn 1962 — Kuhn T.S., *The Structure of Scientific Revolutions*, Chicago 1962, 1970.

Kuhn 1978 — Kuhn T.S., *Black-Body Theory and the Quantum Discontinuity. 1894-1912*, Oxford 1978.

Landau–Liftshitz 1976 — Landau L.D., Liftshitz E. M., *Mechanics*, (Eng. tr.), v. 1 of *The Course of Theoretical Physics*, Oxford 1976.

Lakatos 1978 — Lakatos I., *The Methodology of Scientific Research Programmes*, Cambridge 1978.

Laplace 1820 — Laplace P.-S., *Théorie analytique des Probabilités*, 3rd ed., Paris 1820.

Laroche *et al.* 1995 — Laroche S., Doyère V., Rédini-Del Negro C., Burette F., "Neural Mechanisms of Associative Memory: Role of Long-Term Potentiation", in McGaugh J.L., Weinberger N.M., Lynch G. (Eds.), *Brain and Memory: Modulation and Mediation of Neuroplasticity*, Oxford 1995, pp. 277-302.

Laudan 1977 — Laudan L., *Progress and Its Problems*, Berkeley 1977.

Leibniz 1666 — Leibniz G.W., *Dissertatio de arte combinatoria*, in Leibniz *PS*, v. 4, pp. 27-104.

Leibniz 1678-1679	Leibniz G.W., "Ad ethicam B. d. Sp.", in Leibniz *PS*, v. 1, pp. 139-152.
Leibniz 1684	Leibniz G. W:, "Meditationes de Cognitione, Veritate et Ideis", *Acta Eruditorum Lipsiensium* (November) Repr. in Leibniz *PS*, v. 4, 422-26.
Leibniz 1686a	Leibniz G.W., "Brevis demostratio erroris mirabilis Cartesii et aliorum …, *Acta Eruditorum Lipsiensium* (Repr. in *Mathematische Schriften*, v. 6, 117-123).
Leibniz 1686b	Leibniz G.W., *Discours de Metaphysique*, in Leibniz *PS*, v. 4, 427-63.
Leibniz 1689a	Leibniz G.W., "Marginalia in Newtoni Principia Matematica", (Repr. Paris 1973).
Leibniz 1689b	Leibniz, G.W., "Primæ veritates", in Leibniz *C*, pp. 518-23.
Leibniz 1695	Leibniz G.W., "Systeme nouveau de la nature et de la communication des substances, aussi bien que de l'union qu'il y a entre l'ame et le corps", *Journal des savants*, Repr. in *PS*, v. 4, pp. 477-487.
Leibniz 1702	Leibniz G.W., "Extrait du Dictionnaire de M. Bayle article Rorarius p. 2599 sqq. de l'Edition de l'an 1702 avec mes remarques", in Leibniz *PS*, v. 4, pp. 524-54.
Leibniz 1706-1716	Leibniz G. W., « Briefwechsel zwischen Leibniz und des Bosses », in Leibniz *PS*, v. 2, pp. 285-521.
Leibniz 1710	Leibniz G. W., *Essais de Théodicée*, 1710; in Leibniz *PS*, v. 6, 21-436.
Leibniz 1710-1712	Leibniz G.W., "Principes de la Nature et de la Grace", in Leibniz *PS*, v. 6, pp. 598-606.
Leibniz 1712-1714	Leibniz G.W., "Monadologie", in Leibniz *PS*, v. 6, pp. 607-623.
Leibniz 1765	Leibniz G.W., *Nouveaux Essais sur l'Entendement par l'auteur du système de l'harmonie prestablie*, Amsterdam, 1765; (Repr. in Leibniz *PS*, v. 5, pp. 39-509).
Leibniz *C*	Leibniz G.W., *Opuscules et fragments inédits* (Ed. Couturat), Paris 1903; rep. Hildesheim 1988.
Leibniz *PS*	Leibniz G.W., *Philosophische Schriften* (Ed. Gehardt), Halle 1875; repr. Olms 1978.
Leibniz-Clarke 1715-1716	"Streitschriften Zwischen Leibniz und Clarke 1715-1716", in: Leibniz *PS*, v. 7, pp. 345-440.
Lemaître 1931a	Lemaître G., "Expansion of the universe, The expanding universe", *Monthly Notices of the Royal Astronomical Society* **91**, 490-501.
Lemaître 1931b	Lemaître G., The Beginning of the World from the Point of View on Quantum Theory, *Nature* **127**, 706.
Lessing 1777	Lessing, Gotthold Ephraim, *Die Erziehung des Menschengeschlechts*.
Levi-Strauss 1962	Levi-Strauss C., *Le Pensée Sauvage*, Paris 1962, 1985.
Levi-Strauss 1964	Levi-Strauss C., *Le cru et le cuit*, Paris 1964, 1985.

Levinthal *et al.* 1976 — Levinthal F., Macagno E., Levinthal C., "Anatomy and development of identified cells in isogenic organisms", *Cold Spring Harbour Symp. Quant. Biol.* **40**, 321-331.

Lewis 2009 — Lewis F. A., "Form and Matter", in Anagnostopoulos 2009a, pp. 162-85.

Lewontin 2000 — Lewontin R.C., *The Triple Helix: Gene, Organism, and Environment*, Cambridge (MA).

Libet 2004 — Libet B., *Mind-Time: The Temporal Factor in Consciousness*, Cambridge (MA) 2004.

Lindberg 1992 — Lindberg D.C., *The Beginning of Western Science: The European Scientific Tradition in Philosophical, Religious, and Institutional Context, Prehistory to A.D. 1450*, Chicago 1992, 2007.

Lloyd Morgan 1891 — Lloyd Morgan C., *Animal Life and Intelligence*, Boston 1891.

Lloyd Morgan 1923 — Lloyd Morgan C., *Emergent Evolution*, London 1923, 1927.

Locke 1689 — Locke J., *An Essay Concerning Human Understanding*, 1689; Oxford 1975, 1979, 1987, 1990.

Lonergan 1957 — Lonergan B., *Insight*, London 1957.

Lonergan 1963 — Lonergan B., "Metaphysics as Horizon", *Gregorianum* **44**, 307-318.

Lyotard 1992 — Lyotard J.F., *The Post-Modern Condition: A Report on Knowledge*, Manchester 1992.

Macagno *et al.* 1973 — Macagno E.R., Levinthal C., Lopresti V., "Structural development of neural connections in isogenic organisms: Variations and similarities in the optic system of *Daphnia magna*", *Proc. Natl. Acad. Sci. USA* **70**, 57-61.

Mach 1883 — Mach E., *Die Mechanik in ihrer Entwickelung histrorich-kritisch dargestellt*, Praga 1883.

Mach 1905 — Mach E., *Erkenntnis und Irrtum*, 1905; 2nd ed. 1906, 5th ed. Leipzig 1926, Darmstadt 1991.

Mahillon–Chandler 1998 — Mahillon J., Chandler M., "Insertion Sequences", *Microbiology and Molecular Biology Reviews* **62**, 725-774.

Maimonides GE — Maimonides M., *Le Guide des égarés*, Verdier, 1979.

Maiocchi 1995 — Maiocchi R, *Storia della scienza in occidente. Dalle origini alla bomba atomica*, Firenze, 1995; Milano 2000.

Maldamé 2003 — Maldamé J.-M., *Science et foi en quête d'unité*, Paris 2003.

Maldamé 2011 — Maldamé J.-M., "The Various Meanings of the Word Evolution in Science, Philosophy and Theology" in Auletta *et al.* 2011c, pp. 557-571.

Malebranche 1674-1678 — Malebranche N., *De la recherche de la verité*, Paris 1674-78; 6th ed.: Paris 1712; rep. in *Œuvres* (Ed. G. Rodis-Lewis), Paris, Pleiade, 1979, I, pp. 1-1126.

Maréchal 1926 — Maréchal J., *Le point de départ de la métaphysique. Cahier V,* Louvain-Paris 1926.

Margenau 1950 — Margenau H., *The nature of Physical Reality: A Philosophy of Modern Physics,* New York 1950; Woodbridge Conn. 1977.

Margenau 1984 — Margenau H., *The Miracle of Existence,* Woodbridge Conn. 1984.

Margulis 1970 — Margulis L., *Origin of Eukaryotic Cells,* New Haven 1970.

Maritain 1934 — Maritain J., *Distinguer pour unir. Les degrés du savoir,* 1934; 6th ed. 1958; Desclée de Brouwer 1974.

Markman 1990 — Markman E.M., "Constraints Children Place on Word Meanings", *Cognitive Science* **14**, 57–77.

Markman–Wachtel 1998 — Markman E.M., Wachtel G.F., "Children's Use of Mutual Exclusivity to Constrain the Meanings of Words", *Cognitive Psychology* **20**, 121–57.

Mayr 1988 — Mayr E., *Towards a New Philosophy of Biology: Observations of an Evolutionist,* Cambridge (MA) 1988.

McGrath 2001-2003 — McGrath A.E., *A Scientific Theology,* 3 v., Grand Rapids (MI) and Cambridge (UK) 2001-2003.

McGrath 2005 — McGrath A., *Dawkins' God. Genes, Memes and the Meaning of Life,* Oxford 2005.

Merleau-Ponty 1942 — Merleau-Ponty M., *La structure du comportment,* Paris, 1942 2002.

Merleau-Ponty 1945 — Merleau-Ponty M., *Phénoménologie de la perception,* Paris 1945, 2001.

Mewlana *SW* — Mewlana Jalal al-Din Rumi, *Selection from the Great Works,* Konya (Turkey), 2007.

Mill 1865 — Mill J. S., *An Examination of Sir William Hamilton's Philosophy and of the Principal Philosophical Questions Discussed in His Writings,* 1865; 3rd ed. 1866; London, 1979.

Mo *et al.* 2010 — Mo H., van den Bosch F., and White S., *Galaxy Formation and Evolution,* Cambridge 2010.

Monod 1970 — Monod J., *Le hasard et la nécessité,* Paris 1970.

Morowitz 2002 — Morowitz, H. J., *The Emergence of Everything: How the World Became Complex,* Oxford 2002.

Murdoch 1987 — Murdoch D., *Niels Bohr's Philosophy of Physics,* Cambridge 1987.

Murphy–Brown 2007 — Murphy N., Brown W.S., *Did My Neurons Make Me Do It? Philosophical and Neurobiological Perspectives on Moral Responsibility and Free Will,* Oxford 2007.

Murphy–Stoeger 2007 — Murphy N., Stoeger W.R. (Eds.), *Evolution and Emergence,* Oxford 2007.

Nagel 1961 — Nagel E., *The Structure of Science,* London 1961.

Neurath 1931 — Neurath O., "Physikalismus", *Scientia* 1931, 297-303.

Neurath 1932-1933 — Neurath O., "Protokollsätze" *Erkenntnis* **3**, 204-214.

Neurath 1938 — Neurath O., "Unified Science as Encyclopedic Integration ", *International Encyclopedia of Unified Science*, v. 1, Chicago 1938, pp. 1-27.

Newman 1845 — Newman J. H., *An Essay on the Development of Christian Doctrine*, London 1845; 2nd ed. 1878.

Newman 1870 — Newman J. H., *An Essay in Aid of a Grammar of Assent*, London, 1870.

Newton 1687 — Newton I., *Philosophiae Naturalis Principia Mathematica*, London, 1687; Cambridge, 1713; London, 1726; Harvard, 1972 (Eds. A. Koyré and I. B. Cohen).

Newton 1704 — Newton I., *Opticks or A Treatise of the Reflections, Refractions, Inflections and Colours of Light*, 1704, 2nd ed. lat. 1706, 3rd ed. 1721, London, 4th ed. 1730; New York, 1952, 1979.

Nielsen–Chuang 2000 — Nielsen M.A., Chuang I.L., *Quantum Computation and Quantum Information*, Cambridge, 2000, 2002.

Numbers 2003 — Numbers R.L., "Science without God: Natural Laws and Christian Belief", in: Lindberg D.C., Numbers R. (Eds.), *When Science and Christianity Meet*, Chicago 2003, pp. 265-285.

Numbers 2011 — Numbers R.L., "Scientific Creationism and Intelligent Design" in Auletta *et al.* 2011c, pp. 517-535.

Odum–Odum 1976 — Odum H.T., Odum E.P., *Energy Basis for Man and Nature*, New York 1976.

Otto 1917 — Otto, R., *Das Heilige. Über das Irrationale in der Idee des Göttlichen und sein Verhältnis zum Rationalen.*

Paley 1802 — Paley, W., *Natural Theology: Or, Evidences of the Existence and Attributes of the Deity, Collected from the Appearances of Nature*, 1802; London 1836.

Pannenberg 1993 — Pannenberg W., *Toward a Theology of Nature: Essays on Science and Faith*, Louisville (KY) 1993.

Pannenberg 2008 — Pannenberg W., *The Historicity of Nature: Essays on Science and Theology*, Gregersen N.H. (Ed), West Conshohocken (PA) 2008.

Pangallo 1987 — Pangallo M., *L'Essere come Atto nel Tomismo essenziale di Cornelio Fabro*, Roma 1987

Pangallo 1991 — Pangallo M., *Il principio di causalità nella metafisica di S. Tommaso*, Roma 1991.

Parisi 1999 — Parisi G., "Complex Systems: A Physicist's Viewpoint", *Physica* **A263**, 557-64.

Parisi 2006 — Parisi G., "Complex Systems: A Physicist Viewpoint", in Auletta G. (Ed.), *The Controversial Relationships Between Science and Philosophy: a Critical Assessment*, Vatican City 2006, pp. 85-95.

Pasnau 2004   Pasnau R., "Form, Substance, and Mechanism", *Philosophical Review* **113**, 31-88.

Patzig 1960   Patzig G., "Theologie und Ontologie in der Metaphysik des Aristoteles", *Kant—Studien* **52**, 185—205.

Pawlowski *et al.* 2009   Pawlowski M., Paterek T., Kaszlikowski D., Scarani V., Winter A., Zukowski M.Z., "Information Causality as a Physical Principle", *Nature* **461**, 1101-1104.

Paz *et al.* 1993   Paz J.P., Habib S., Zurek W.H., "Reduction of the wave packet: Preferred observable and decoherence time scale", *Physical Review* **D47**, 488-501.

Peacocke 1993   Peacocke A. *The Theology for Scientific Age: Being and Becoming – Natural, Divine and Human*, Minneapolis 1993.

Pedersen 2007   Pedersen O., *The Two Books: Historical Notes on some Interactions between Natural Science and Theology*, Vatican City 2007.

Peirce 1866   Peirce C.S., *The Logic of Science or Induction and Hypothesis: Lowell Lectures*, in Peirce *W*, I, pp. 357-504.

Peirce 1868   Peirce C.S., "Some Consequences of Four Incapacities", *Journal of Speculative Philosophy* **2** 140–57 (Repr. in Peirce *W*, II, pp. 211-42).

Peirce 1869   Peirce C.S., "Grounds of Validity of the Laws of Logic: Further Consequences of Four Incapacities", *Journal of Speculative Philosophy* **2**, 193–208 (Repr. in Peirce *W*, II, pp. 242-272).

Peirce 1871   Peirce, C.S., "Fraser's *The Works of George Berkeley*", *North American Review* **113**, 449–72 (Repr. in Peirce *W*, II, pp. 462-487).

Peirce 1877   Peirce C.S., "The Fixation of Belief", *Popular Science Monthly* **12**: 1-15 (Repr. In Peirce *W*, III, pp. 242-257).

Peirce 1878a   Peirce C.S., "Deduction, Induction, and Hypothesis", *Popular Science Monthly* **13**, 470-82 (Repr. in Peirce *W*, III, pp. 323-338).

Peirce 1878b   Peirce C.S., "How to Make Our Ideas Clear", *Popular Science Monthly* **12**, 286-302 (Repr. in Peirce *W*, III, pp. 257-276).

Peirce 1887-1888   Peirce C.S., *A Guess at the Riddle*, in Pierce *W*, VI, pp. 165-210.

Peirce 1888   Peirce C.S., "Trichotomic", in Peirce *W*, VI, pp. 211-215.

Peirce 1891   Peirce C.S., "The architecture of theories", *The Monist* **1**, 161-176. (Repr. in *EP*, I, pp. 285-297).

Peirce 1892a   Peirce C.S., "The doctrine of necessity examined", *The Monist* **2**, 321-337 (Repr. in Peirce *EP*, I, pp. 298-311).

Peirce 1892b   Peirce C.S., "The law of mind", *The Monist* **2**, 533-59 (Repr. in Peirce *EP*, I, pp. 312-333).

Peirce 1893   Peirce C.S., "Evolutionary Love", *The Monist* **3**, 176-200 (Repr. in Peirce *EP*, I, pp. 352-371).

Peirce 1898a   Peirce C. S., "Philosophy and the Conduct of Life", in Peirce *EP*, II, pp. 27-41.

Peirce 1898b	Peirce C.S., "The First Rule of Logic", in Peirce *EP*, II, pp. 42-56.
Peirce 1901	Peirce C.S., "On the Logic of Drawing History from Ancient Documents, Especially from Testimonies, in Pierce *EP*, II, pp. 75-114.
Peirce 1903a	Peirce C.S., "The Maxim of Pragmatism", in Pierce *EP*, II, pp. 133-144.
Peirce 1903b	Peirce C.S., "Pragmatism and the Logic of Abduction", in Peirce *EP*, II, pp. 226–41.
Peirce 1905a	Peirce C.S., "What Pragmatism Is", *Monist* **15**, 161-181 (Rep. in Pierce *EP*, II, pp. 331-345).
Peirce 1905b	Peirce C.S., "Issues of Pragmatism", *Monist* **15**, 481-99 (Rep. in Pierce *EP*, II, pp. 346-359).
Peirce 1906	Peirce C.S., "The Basis of Pragmaticism in Phaneroscopy", in Pierce *EP*, II, pp. 360-370.
Peirce *CP*	Peirce C.S., *The Collected Papers*, Vols. I-VI (Hartshorne C., Weiss P. Ed), Cambridge (MA) 1931-1935; vols. VII-VIII (Burks A.W. Ed), Cambridge (MA) 1958.
Peirce *EP*	Peirce C.S., *The Essential Peirce*, Bloomington 1998, vols I-II.
Peirce *W*	Peirce C.S., *Writings*, Bloomington, 1982-.
Penrose 1989	Penrose R., *The Emperor's New Mind*, Oxford, 1989.
Penrose 2004	Penrose R., *The Road to Reality: A Complete Guide to the Laws of the Universe*, London 2004.
Penrose 2010	Penrose R., *Cycles of Time: An Extraordinary New View of the Universe*, London 2010.
Piaget 1936	Piaget J., *La naissance de l'intelligence chez l'enfant*, 1936, Neuchâtel 1977.
Piaget 1945	Piaget J., *La construction du réel chez l'enfant*, 1945, Neuchâtel 1967.
Planck 1900a	Planck M., "Über die Verbesserung der Wien'schen Spektralgleichung", *Verhandlungen der Deutchen Physikalischen Gesellschaft* **2**, 202-204.
Planck 1900b	Planck M., "Zur Theorie des Gesetzes der Energieverteilung im Normalspektrum", *Verhandlungen der Deutchen Physikalischen Gesellschaft* **2**, 237-45.
Planck 1965	Planck M., *Vorträge und Erinnerungen*, Darmstadt 1965.
Plato *O*	*Platonis opera*, Oxford, 1900, 1987.
Plato, *Parmenides*	Plato "Parmenides", in Plato *O*, b. II.
Plotinus *Enn.*	Plotinus, *Enneads*, (ed. A. Kirchhoff) Leipzig 1856.
Poincaré 1902	Poincaré H., *La science et l'hypothèse*, Paris 1902; Paris 1968.
Polkinghorne 1998	Polkinghorne J., *Belief in God in an Age of Science*, New Haven (CT) 1998.
Polkinghorne 2005	Polkinghorne J., *Exploring Reality: The Intertwining of Science and Religion*, London 2005.

Polkinghorne 2006   Polkinghorne J., *Science and Creation: The Search for Understanding*, West Conshohocken (PA) 2006.

Polkinghorne 2009   Polkinghorne J., *Theology in the Context of Science*, New Haven 2009

Polkinghorne–Beale 2009   Polkinghorne J., Beale N., *Questions of Truth*, Louisville (KY) 2009.

Popper 1934   Popper K.R., *Logik der Forschung*, Wien 1934, 8th Mohr, 1984.

Popper 1963   Popper K.R., *Conjectures and Refutation*, London 1963.

Popper–Eccles 1977   Popper K.R., Eccles J.C., *The Self and Its Brain: An Argument for Interactionism*, Berlin 1977.

Poupard 1983   Poupard P., "Galileo Galilei: 350 Years of Subsequent History", in Poupard P. (Ed.), *Galileo Galilei: Toward a Resolution of 350 Years of Debate - 1633-1983*, Pittsburgh 1987.

Poupard 1987   Poupard P. (Ed.), *Galileo Galilei: Toward a Resolution of 350 Years of Debate – 1633-1983*, Pittsburgh 1987.

Poupard 1992   Address of Cardinal Paul Poupard to the Plenary Session of the Pontifical Academy of Sciences giving a summary of the Conclusions of the Galileo Galilei Commission, October 31, 1992; in *Papal Addresses*, Vatican City 2003, pp. 344-348.

Poupard 1994   Poupard P. (Ed.), *Après Galilée. Science et foi: nouveau dialogue*, Paris.

Putnam 1982   Putnam H., "Three Kinds of Scientific Realism", *The Philosophical Quarterly* **32**, 195-200.

Quine 1948   Quine W.V.O., "On What There Is", *Review of Meta- physics* (1948) (Repr. in Quine 1953, pp. 1-19).

Quine 1951   Quine W.V.O., "Two Dogmas of Empiricism", *The Philosophical Review* **60**: 20-43 (Repr. in Quine W.V.O., *From a Logical Point of View*, Cambridge (MA) 1953).

Quine 1953   Quine W.V.O., *From a Logical Point of View*, Cambridge (MA) 1953, 2nd ed. 1961; 1980.

Quine 1960   Quine W.V.O., *Word and Object*, Cambridge (MA) 1960, 1989.

Quine 1969   Quine W.V.O, *Ontological Relativity and Other Essays*, New York 1969.

Rahner 1976   Rahner K., *Grundkurs des Glaubens. Einführung in den Begriff des Christentums*, Freiburg 1976, 1984.

Ray 1727   Ray J., *The Wisdom of God Manifested in the Works of the Creation*, London 1727.

Rees 1999   Rees, Martin, *Just Six Numbers*, 1999; London, 2000.

Reichenbach 1928   Reichenbach H., *Philosophie der Raum-Zeit-Lehre*, Berlin 1928.

Reichenbach 1944   Reichenbach H., *Philosophic Foundations of Quantum Mechanics*, Berkeley 1944.

Reid 1764   Reid T., *An Inquiry into the Human Mind on the Principles of Common Sense*, 1764; 5th ed. Edinburgh 1801.

Renfrew 2011 — Renfrew A.C., "The concept of evolution as applied to the development of human cultures",in Auletta *et al.* 2011c, pp. 307-314.

Rosenzweig 1921 — Rosenzweig F., *Der Stern der Erlösung*, 1921, 1930, 1976, Frankfurt am Main 1988.

Rossi 1973-1974 — Rossi A., "Le due strade della fisica", *Scientia*, issue July-August 1973 and issue IX-XII 1974.

Ruini 2010 — Ruini C., "Le vie di Dio nella ragione contemporanea", in *Dio oggi. Con Lui o senza di Lui cambia tutto*, Siena 2010, pp. 29-56.

Russell 1902 — Russell B., "Letter to Frege," in van Heijenoort J., *From Frege to Gödel*, Cambridge (MA) 1967, pp. 124-125.

Russell 1903 — Russell B., *Principles of Mathematics*, London 1903 (Routlegde 1992).

Russell 1931 — Russell B., *The Scientific Outlook*, London 1931.

Russell 2006 — Russell R.J., *Cosmology, Evolution and the Resurrection Hope: Theology and Science in Creative Mutual Interaction*, Kitchener 2006.

Russell 2008 — Russell R.J., *Cosmology: From Alpha to Omega. The Creative Mutual Interaction of Theology and Science*, Minneapolis 2008.

Russell *et al.* 1988 — Russell R.J., Stoeger W.R., Coyne G.V. (Ed), *Physics, Philosophy, and Theology: A Common Quest for Understanding*, Vatican City and Berkeley (CA) 1988, 1995, 2000.

Russell *et al.* 1995 — Russell R.J., Murphy N., Peacocke A.R. (Ed.), *Chaos and Complexity: Scientific Perspectives on Divine Action*, Vatican City and Berkeley (CA) 1995, 1997, 2000.

Russell *et al.* 1998 — Russell R.J., Stoeger W.R. Ayala, F.J. (Eds.), *Evolutionary and Molecular Biology: Scientific Perspectives on Divine Action*, Vatican City and Berkeley (CA) 1998.

Russell *et al.* 2008 — Russell R.J., Murphy N., Stoeger W.R. (Eds.), *Scientific Perspectives on Divine Action: Twenty Years of Challenge and Progress*, Vatican City and Berkeley (CA) 2008.

Salucci 2011 — Salucci A., *"In principio…" Variazioni sul tema della Creazione*, Vatican City 2011.

Saunders 2002 — Saunders N., *Divine Action and Modern Science*, Cambridge, 2002.

Schlichting and Pigliucci 1998 — Schlichting, C. D. and Pigliucci, M., *Phenotypic Evolution: A Reaction Norm Perspective*, Sunderland (MA) 1998.

Schrödinger 1935 — Schrödinger E., "Die gegenwärtige Situation in der Quantenmechanick" I-III, *Naturwissenschaften* **23**, 807-812, 823-828, 844-849.

Schrödinger 1944 — Schrödinger E., *What is Life?*, in Schrödinger 1992, pp. 1-90.

Schrödinger 1958 — Schrödinger E., *Mind and Matter*, Cambridge 1958 (Repr. in Schrödinger 1992, pp. 91-164).

Schrödinger 1992 — Schrödinger E., *What is Life? with Mind and Matter and Autobiographical Sketches*, Cambridge 1992.

Sellars 1920 — Sellars R.W., *Critical Realism: A Study of the Nature and Conditions of Knowledge*, London 1920.

Sellars 1922 — Sellars R.W., *Evolutionary Naturalism*, Chicago 1922.

Selvaggi 1964 — Selvaggi F., *Causalità e indeterminismo. La problematica moderna alla luce della filosofia aristotelico-tomista*, Rome 1964.

Selvaggi 1985 — Selvaggi F., *Filosofia del mondo. Cosmologia filosofica*, Rome 1985; 2nd ed. 1993.

Shannon 1948 — Shannon C.E., "A Mathematical Theory of Communication", *Bell System Technical Journal* 27, 379-423, 623-656.

Shimony 1989 — Shimony A., "Search for a Worldview which Can Accommodate our Knowledge of Microphysics", in Cushing J., McMullin E. (eds.), *Philosophical Consequences in Quantum Theory: Reflections on Bell's Theorem*, Notre Dame 1989, pp. 25-37.

Silva 2010 — Silva I. A., *Divine Action in Nature: Thomas Aquinas and the Contemporary Debate*, PhD dissertation, University of Oxford.

Sklar 1974 — Sklar L., *Time and Spacetime*, Berkeley 1974.

Smart 1954 — Smart, J. J. C., "The Temporal Asymmetry of the World", *Analysis* 14, 79-83.

Snow 1960 — Snow C.P., *The Two Cultures*, Cambridge 1960.

Sorge 2006 — Sorge B., *Introduzione alla dottrina sociale della Chiesa*, Brescia 2006; 2nd ed. 2011.

Spencer 1855 — Spencer H., *Principles of Psychology*, 1855, New York 1896, 1920.

Spencer 1860-1862 — Spencer H., *First Principles*, 1860-1862, 6th ed. 1900; London, 1937; 3rd rep. 1946.

Spinoza 1677 — Spinoza B., *Ethica*, Amsterdam 1677.

Stoeger 1995 — Stoeger W.R., "Describing God's Action in the World in Light of Scientific Knowledge of Reality", in Russell *et al.* 1995, pp. 239-261.

Stoeger 1998 — Stoeger W., "The Immanent Directionality of the Evolutionary Process, and Its Relationship to Teleology", in Russell *et al.* 1998, pp. 163-190.

Stoeger 2007 — Stoeger W.R., "Reductionism and Emergence: Implications for the Interaction of Theology with the Natural Sciences", in Murphy–Stoeger 2007.

Suárez 1597 — Suárez F., *Disputationes metaphysicae*, Salamanca 1597.

Tanzella-Nitti 1994 — Tanzella-Nitti G., "Origins, Time and Complexity: A Comment on the Relation between Christian Theology of Creation and Contemporary Cosmology", in Coyne G., Schmitz-Moormann K. (Eds.), *Origins, Time, Complexity*, Geneva 1994, vol. II, 26-36.

Tanzella-Nitti 1997 — Tanzella-Nitti G., "The Aristotelian-Thomistic Concept of Nature and the Contemporary Debate on the Meaning of Natural Laws", *Acta Philosophica* 6, 237-264.

Tanzella-Nitti 2001 Tanzella-Nitti G., "The Book of Nature and the God of Scientists according to the Encyclical *Fides et Ratio*", in *The Human Search for Truth: Philosophy, Science and Faith. The Outlook of the Third Millennium*, Philadelphia 2001, pp. 82-90.

Tanzella-Nitti 2002 Tanzella-Nitti G., "Creazione", in Tanzella-Nitti G., Strumia A. (Ed), *Dizionario interdisciplinare Scienza e Fede. Cultura scientifica, Filosofia e Teologia*, Vatican City 2002, pp. 300-321.

Tanzella-Nitti 2004 Tanzella-Nitti G., "L'ontologia di Tommaso d'Aquino e le scienze naturali", *Acta Philosophica* **13**, pp. 137-155.

Tanzella-Nitti 2008 Tanzella-Nitti G., "The Influence of Scientific World View on Theology: A Brief Assessment and Future Perspectives" in Auletta G. (Ed), *The Relationships Between Science and Philosophy: New Opportunities for a Fruitful Dialogue*, Vatican City 2008, pp. 131-154.

Tanzella-Nitti 2009 Tanzella-Nitti G., *Faith, Reason and the Natural Sciences. The Challenge of the Natural Sciences in the Work of Theologians*, Aurora (CO) 2009.

Tanzella-Nitti 2010 Tanzella-Nitti G., "The Influence Scientific World View on Theology: A Brief Assessment and Future Perspectives", in Desbouts C., Mantovani M. (Eds.), *Didattica delle scienze*, Vatican City 2010, pp. 139-163.

Tarozzi–Colagè 2009 Tarozzi G., Colagè I., "Realismo scientifico e realismo empirico: è possibile discriminare sperimentalmente nel caso dell'interpretazione della meccanica quantistica?", in Alai M. (Ed), *Il realismo scientifico di Evandro Agazzi*, Urbino 2009, pp. 131-156.

Taylor 1957 Taylor R., "The Problem of Future Contingencies" , *Philosophical Review* **66**, 1-28.

Taylor 1998 Taylor R. C., "Aquinas, the Plotiniana Arabica, and the Metaphysics of Being and Actuality", *Journal of the History of Ideas* **59**, 217-239.

Teilhard de Chardin 1923 Teilhard de Chardin P., *La messe sur le monde*, Paris 1923.

Teilhard de Chardin 1955 Teilhard de Chardin P., *Le phénomène humaine*, Paris 1955.

Theunissen 1980 Theunissen M., *Sein und Schein. Die kritische Funktion der Hegelschen Logik*, Frankfurt a. M 1980.

Thirring 2007 Thirring W., *Cosmic Impressions: Traces of God in the laws of Nature*, Philadelphia-London 2007

Thorne 1994 Thorne, K. S., *Black Holes and Time Warps: Einstein's Outrageous Legacy*, New York 1994.

Tillich 1951-1963 Tillich P., *Systematic Theology*, Chicago 1951-1963.

Tindal 1730 Tindal M., *Christianity as Hold as the Creation or the Gospel, a Republication of the Religion of Nature*, London 1730.

Todisco 1999 Todisco O. *Averroè nel dibattito medievale*, Milano 1999.

Tolman 1932 — Tolman E.C., *Purposive Behavior in Animals and Men*, New York 1932.

Torrance 1969 — Torrance T.F., *Space, Time and Incarnation*, Edinburgh 1969.

Torrance 1985a — Torrance T.F., *Reality and Scientific Theology*, Edinburgh 1985.

Torrance 1985b — Torrance T.F., *The Christian Frame of Mind*, Edinburgh 1985.

Torretti 1983 — Torretti R., *Relativity and Geometry*, New York 1983.

Tryon 1973 — Tryon E.P., "Is the universe a vacuum fluctuation?", *Nature* **246**, 396-397.

Ulanowicz 1986 — Ulanowicz R.E., *Growth and Developmant: Ecosystems Phenomenology*, New York 1986.

Vadet 1976 — Vadet J.-C., "Le Primat de l'Action sur l'Etre et l'Essence fil conducteur de la logique d' Al-As'ari?", *Studia islamica* **44**: 25-60.

Van der Leeuw 1933 — Van der Leeuw G., *Phänomenologie der Religion*, 1933; 2nd edition. Tübingen 1956.

Vedral 2010 — Vedral, V., *Decoding Reality: The Universe as Quantum Information*, Oxford 2010.

Verlinde 2011 — Verlinde E., "On the origin of gravity and the laws of Newton", ArXiv, 1001.0785v1.

Von Helmholtz 1867 — von Helmholtz H. L. F., *Handbuch der physiologischen Optik*, 1867; 2nd ed.; Hamburg and Leipzig, 1896.

Von Neumann 1932 — von Neumann, J., *Mathematische Grundlagen der Quantenmechanik*, Berlin 1932.

Von Neumann 1955 — von Neumann, J., *Mathematical Foundations of Quantum Mechanics*, Princeton (Eng. trans. of von Neuman 1932).

Wagner 2000 — Wagner P.J., "Exhaustion of morphological character states among fossil taxa", *Evolution* **54**, 365-386.

Wagner 2005 — Wagner, A., *Robustness and Evolvability in Living Systems*, Princeton 2005.

Watson–Crick 1953 — Watson J.D., Crick F.M., "Molecular structure of nucleic acids", *Nature* **171**, 737-738.

Wedin 2009 — Wedin M. V., "The Science and Axioms of Being", in Anagnostopoulos 2009a, pp. 125-43.

West-Eberhard 2003 — West-Eberhard, M. J., *Developmental Plasticity and Evolution*, Oxford 2003.

Weyl 1949 — Weyl E., *Philosophy of Mathematics and Natural Sciences*, Princeton 1949.

Wheeler 1983 — Wheeler J.A., "Law Without Law", in Wheeler–Zurek 1983, pp. 182–213.

Wheeler 1990 — Wheeler J.A., "Information, Physics, Quantum: The search for Links", in Zurek 1990, pp. 3-28.

Wheeler–Zurek 1983   Wheeler J. A., Zurek W. (Eds.), *Quantum Theory and Measurement*, Princeton 1983.

White 1896   White A. D., *A History of the Warfare of Science with Theology in Christendom*, 1896.

Whitehead 1929   Whitehead A.N., *Process and Reality*, London-New York 1929; corrected ed. 1978, 1979.

Wigner 1960   Wigner E., "The Unreasonable Effectiveness of Mathematics in the Natural Sciences", *Communications in Pure and Applied Mathematics* **13**, 1-13.

Wigner 1961   Wigner E.P., "Remarks on the Mind-Body Question", in I. J. Good (Ed.), *The Scientist Speculates*, London 1961, pp. 284-302.

Wilkins 2002   Wilkins A.S., *The Evolution of Developmental Pathways*, Sunderland (MA) 2002.

Wilson 1998   Wilson E.O., *Consilience: The Unity of Knowledge*, New York 1998.

Wittgenstein 1953   Wittgenstein L., *Philosophische Untersuchungen*, Oxford 1953.

Wolberg *et al.* 1991   Wolberg C., Vershon A.K., Liu B., Johnson A.D., Pabo C.O., "Crystal structure of a MATa2 homeodomain-operator complex suggests a general model for homeodomain-DNA interactions", *Cell* **67**, 517-528.

Wolff 1730   Wolff C., *Philosophia prima sive Ontologia*, Frankfurt a. M. 1730; 2nd ed. 1736.

Wolff 1731   Wolff C., *Cosmologia generalis*, Frankfurt a. M. 1731; 2d ed. 1737.

Wolff 1736   Wolff C., *Theologia generalis*, pars I, Frankfurt a. M. 1736; 2d ed. 1739.

Wolff 1737   Wolff C., *Theologia generalis*, pars II, Frankfurt a. M. 1737; 2nd ed. 1741.

Wright 2007   Wright T., *Surprised by Hope*, London 2007.

Zurek 1990   Zurek W.H. (Ed.), *Complexity, Entropy and the Physics of Information*, Redwood City 1990.

Zwiebach 2004   Zwiebach B., *A First Course in String Theory*, Cambridge 2004.

Zycinski 2006a   Zycinski J.M., *God and Evolution. Fundamental Questions of Christian Evolutionism*, Washington 2006.

Zycinski 2006b   Zycinski J.M., "Evolution and Christian Thought in Dialogue according to the Teaching of John Paul II", *Logos: A Journal of Catholic Thought and Culture* **9**, 13-27.

Zycinski 2006c   Zycinski J.M., "The Laws of Nature and Plato's Theory of Forms", in Auletta G. (Ed), *The Controversial Relationships Between Science and Philosophy: a Critical Assessment*, Vatican City 2008, pp. 175-185.

Ivan Colagè, Ph.D, has conducted his studies in philosophy at the Pontifical Gregorian University (Rome), taking his Doctorate within the Specialization in Science and Philosophy, under the guidance of Professor Gennaro Auletta. He presently teaches a course on the scientific truth at the Faculty of Philosophy of the Pontifical Gregorian University. For "Analecta Gregoriana", he has published, in 2010, his doctoral dissertation *Interazione e inferenza. Epistemologia scientifica ispirata al Pensiero di C. S. Peirce* (Interaction and inference: scientific epistemology inspired by the thought of C. S. Peirce).

Paolo D'Ambrosio is ABD doctoral student in philosophy at the Pontifical Gregorian University under the guidance of Professor Gennaro Auletta. His research is focussed on philosophy of biology. He was a member of the Organizing Committee of the Conference *Biological Evolution: Facts and Theories* held at the Pontifical Gregorian University in 2009. He participated to the realization of the volume collecting the proceedings of the Conference edited by G. Auletta, M. Leclerc, R. Martínez and published by the Gregorian & Biblical Press in 2011 ("Analecta Gregoriana" 312).

Lluc Torcal, Prior of Cistercian Abbey of Santa Maria de Poblet (Catalonia), has a License in Physics by the Univesitat Autònoma de Barcelona, and took a License in Philosophy and a Bachelor's degree in Theology at the Pontifical Gregorian University (Rome). He is currently writing his PhD dissertation in Philosophy under the guidance of Professor Gennaro Auletta in a work about the interpretation of quantum mechanics. Recently, together with Professor Auletta, he has published the paper "From Wave-Particle to Features-Event Complementarity" in the *International Journal of Theoretical Physics*. He is author of several other articles about the relationship among Theology, Philosophy and Sciences.